THE POISONS

AROUND US

THE POISONS
AROUND US

Toxic Metals in Food, Air, and Water

Henry A. Schroeder, M.D.

INDIANA UNIVERSITY PRESS

Bloomington   London

Published in Canada by Fitzhenry & Whiteside Limited, Don Mills, Ontario
Manufactured in the United States of America

Library of Congress Cataloging in Publication Data

Schroeder, Henry Alfred, 1906–
    The poisons around us.

    Bibliography: p. 139
    1. Pollution. 2. Metals. 3. Metals—Physiological
effect. I. Title. [DNLM: 1. Air pollution—
Popular works. 2. Metals—Poisoning. 3. Popular
works. 4. Water pollution—Popular works. QV290
S381p 1974]
TD177.S37 1974 615.9'25'3 73–15283
ISBN 0–253–16675–6

TO DAVID,

one of the best

CONTENTS

Preface, xi

I Too Much is Too Much, 1
II Metals and the Human Body, 6
III Metals and Civilization, 15
IV Environmental Contamination by Nature and Industry, 26
V Lead, the Lead Industry, and Man's Health, 36
VI Mercury as a Cause of National Paranoia, 59
VII Cadmium, the Dragon's Teeth, 73
VIII Elements Toxic and Not So Toxic, 93
IX Pure Food is Poor Food, 111
X Control of Metal Pollution, 130

Bibliography, 139
Index, 141

TABLES

III-1 Elements Essential for Life or Health, 22

III-2 Toxicity of Important Elements, 24

IV-1 Essential Elements in Seawater and in Air, 28

IV-2 Toxic and Inert Elements in Seawater and in Air, 30

IV-3 Potential Pollution of Fresh Water and Seawater, 32

V-1 Airborne Lead, 42

V-2 Some Sources of Lead Pollution, 51

V-3 Lead in Street Dirt of 77 American Cities, 55

V-4 Incidence of Overexposure to Lead, 56

V-5 Sources of Lead Other than Leaded Gasoline, 57

VI-1 Sources of Exposure to Mercury, 61

VI-2 Mercury in the Natural Environment, 62

VI-3 Effects of Industrial Discharges of Mercury, 66

VII-1 Metals Found in 720 Rivers and Lakes of the U.S., 76

VII-2 Cadmium Concentrations in Fish, 78

VII-3 Effects of Cadmium in Food and Water, 82

VII-4 The Role of Drinking Water in Deaths from Heart Diseases, 88

VII-5 Potential Exposures of Human Beings to Cadmium, 89

VII-6 Cadmium in Human Livers and Kidneys, 91

VIII-1 Potential Human Exposures to Toxic Trace Metals, 94

VIII-2 Metals in Human Hair, 101

VIII-3 Diseases and Disorders in Workers Exposed to Metals, 106

VIII-4 Principal Sources of Human Exposures
to Toxic Trace Metals, 108

VIII-5 Economic Losses from Industry and Weathering, 109

ix-1 Amounts of Essential Trace Elements
      in 2-Day Hospital Diets, 118
ix-2 Types of Food Containing Essential Micronutrients, 119
ix-3 Nutrients Lost in Refining Whole Products, 121
ix-4 Metal Concentrations of Alcoholic Beverages, 123
ix-5 Compensating for Empty Calories, 124
ix-6 Recommended Daily Allowances of Vitamins and Minerals, 128
x-1   Limits on Concentrations of Airborne Metals, 135
x-2   Limits on Concentrations of Waterborne Metals, 136

PREFACE

In the last decade many books have
been written about pollution, and the literature is filled with thousands
of articles on the subject. Pollution is one of the most popular themes
for retired scientists, geographers, science writers, plant biologists, and
environmentalists (a new breed of cat) to expound on, and ecology
has become a household word. Most of these writers are pessimists,
and predict the end of the world in half a century. A few of them
are optimistic about our future if something is done now. A rare one
is imaginatively pragmatic, predicting a dirty, uncomfortable world
of short-lived, undernourished people, malformed by birth defects and
crowding the planet—but not the end of all life on earth.

These Cassandras make illogical jumps from reality to fantasy.
Pollutants are bad for us. They look bad, they smell bad, they taste
bad, they feel bad, they sound bad. They may do harm to the environment, making some birds extinct, denuding places of vegetation,
but by no stretch of the imagination will they become lethal to all
mankind. If man does not control them, natural law will, for they
are a function of population and they increase as the population ex-
plodes and decrease as it declines or as the people act. Nowhere is
there the pollution of eighteenth-century London, for example.

All organic substances are eventually biodegradable, except the
great class of plastics. Even if man covers the earth with paper, sewage, garbage, rubber, and wood products, bacteria and molds will

slowly but inevitably decay them and they will go back to the elements from which they were made. Even the organic pesticides are bio-degradable.

Not so the metals. No metal—or element—is biodegradable. If released into the environment, all metals will accumulate until they are leached out of soil to enter the sea. In the sea they tend to fall to the bottom. If in the process they enter the body of man, they may do good or harm. It is very hard to get rid of them. Too little attention has been paid to them. For this reason they are important.

Man in his infinite wisdom and in his evolution towards civilization has seen fit to dig up metals from deposits and scatter them over the face of the globe. That some of them are toxic has worried him little in the past. But now he is slowly becoming concerned.

This book discusses metals and other elements as pollutants and as substances beneficial to life. It is concerned primarily with human health, and thus it places pollutants in their proper perspectives. It is based on a knowledge of metals, their toxicity (especially the use of low-level doses of metals given for a lifetime), and disease.

Source material on which this book is based includes publications of the Trace Element Laboratory of Dartmouth Medical School, some hundred of them during the past twenty years; *Trace Elements in Biochemistry* by H. J. M. Bowen; *Trace Elements in Human and Animal Nutrition* by E. J. Underwood; and Chapter II by Isabel H. Tipton and Chapter III by Gwyneth Parry Howell in *Report of the Task Group on Reference Man,* a report of the International Commission on Radiological Protection, Committee 2, in preparation. The author is indebted to all of his colleagues and technicians who have contributed knowledge of the subject, to the National Institutes of Health (from whom all blessings flow—or did flow for 28 years) for support of the various works, to Professor S. Marsh Tenney, who has lent invaluable moral support through full and lean years, and to Mrs. Margery Burrill for patience in typing the manuscript.

Ginseng Hill,
Brattleboro, Vermont

Tree Tops,
St. Thomas, Virgin Islands
1972–1973

THE POISONS

AROUND US

TOO MUCH IS TOO MUCH

Some people think that man has so fouled his own nest—the earth—that he will have killed himself off from his own wastes in another twenty-five or fifty years.

This dire end to mankind could not happen. As long as man reacts to threats by counteraction, he will control the threats more or less effectively. When he knows what the problem is, he will solve it.

Most people do not take this doomsday view of the future. Many have a fear, however, that the modern environment is bad for us in some vague and subtle way. Contaminants are everywhere. There are poisons in our food which shorten our lives or give us deformed babies, cancer, and other terrible ills, they say. The air chokes on bad days and is still toxic on good days. Our water supplies contain unknown wastes which make us sick. So they say. And worst of all is the unseen menace of radioactivity, hanging over us like the sword of Damocles.

So they take evasive action. They go to health food stores, to pay high prices for simple or exotic foods, termed "organic." They drink distilled water, or bottled water, or mineral water. They move to the country when they can, to avoid the bad air. They violently oppose nuclear power plants (which do not dirty the air). They live in a state of constant low-key fear. Is there reason for this fear?

Animals take action by fleeing or fighting. The human animal also reacts by using logic based on knowledge. He learns all he can about the problem, discovers its sources, and solves it by controlling it.

Recognition of the problem as a problem, research to discover its causes, and application of knowledge to prevent the problem from developing is the only logical way of dealing with it.

For example, for centuries many infants, some children, and a sizeable number of adults have died of intestinal infections—diarrhea, dysentery, typhoid, cholera, and the like. But not until the discovery of bacteria during the latter half of the nineteenth century were the causes slowly revealed. The excreta of human beings carrying or suffering from these diseases were polluting the environment and, with it, food and water which other people swallowed. Not a very pleasant idea, but that is what is still happening today.

Flies can carry the organisms from feces to food; so can fingers. Human sewage can pollute drinking water. Shellfish can extract the organisms from seawater, and be eaten raw. Every case of "Montezuma's Revenge," "Gippy Tummy," or "Tourista" comes from food or water contaminated by human feces. So do cases of infectious hepatitis from shellfish; viruses have lately been added to the list of fecal-borne pathogens.

The logical steps for control were developed: chlorination of water supplies, which kills bacteria and, with vaccines, has virtually eliminated typhoid fever and dysentery in our cities; sewage treatment plants, in which good bacteria destroy bad; septic tanks, which are small treatment plants; hygiene enforced by law in restaurants; little bits of control everywhere. During the past thirty years, with the advent of antibiotics, failures of the control systems could be treated by curing the disease when it occurred to the point where what were once the most frequent causes of death are now among the least. Formerly only the immune survived; now the susceptible survive along with the immune. Curing the disease has become less important than preventing it.

Although diseases transmitted from man to man via human feces are not completely controlled, and those caused by tough viruses are only partly controlled (because the viruses live through sewage treatment plants), the methods are known and can be applied, given adequate facilities. All pollutants which cause human diseases are capable of control, granted that the diseases are recognized and the cause discovered. Among these pollutants are the metals.

Metal pollution of the environment and of man is as old as metallurgy—some six thousand years or more, although its effects on human beings are only being recognized today. In order to control the diseases caused by metals and other elements, we must examine each element separately and carefully in the light of its biological effects, test the idea that a little bit given for a lifetime slowly poisons us so that we don't live as long or as healthily as we should, and decide whether or not an element is good for us (like iron), does nothing to us (like aluminum), or accumulates in our bodies to do us eventual harm (like lead). We must look at sources of pollution—exposures to humans in air, food, and water—and the rate of increase in exposure so as to foretell eventual hazards in the near or distant future. Control comes naturally from knowledge of this sort, just as control of fecal-borne diseases was achieved, step by step. Today, enough knowledge is available to outline a logical program for control of metal pollution in relation to its importance and to the toxic effects of the metals around us.

Extensive research has revealed a law of Nature which can be called "The Toxicity of Relative Elemental Abundance." Elements which are highly abundant on the earth and in seawater are relatively nontoxic. Elements which are scarce on the earth and in seawater are usually toxic to living things. Toxicity of elements varies inversely with abundance.

This law is generally true, although there are always exceptions. No toxic metals, however, are abundant on the earth's surface. The law is logical, for life developed and evolved in a constant milieu of elements of varying toxicity. The law is meaningful, for man has chosen to dig from the earth vast quantities of scarce metals and spread them across the face of the globe, in some cases doing himself harm.

It is necessary, in fact mandatory, for us to avoid the toxic elements in air, food, and water. It would be silly and stupid for us to exert great efforts to avoid elements which do us no harm or do us good. This book compiles briefly what we know about the elements which might be harmful to us, and discloses those which are inert or are necessary for health.

To every change there is resistance by the vested interests. In the cases of pollution by elements, some of the interests are powerful, others small and concentrated, but all will proclaim anything in the least bit believable, true or false (or outright lie), in order to protect themselves from invasion and loss of profits. Some of these Machiavellian machinations have become known to those who have been their object.

The subtle effects of traces of pollutants, be they living bacteria and viruses, nutritive chemicals, or toxic elements, may be masked by the obvious and visible changes induced in air and water. Raw sewage is highly nutritive and causes overgrowth of algae in stagnant lakes, crowding out other life. There is generally less of this odorous process now than a hundred years ago, when the Houses of Parliament had to close because of the stench from the Thames. We control sewage now in most places. The brown pall which hangs over the industrial heartlands and seaboards is made of dust and smoke, which screen out some of the solar energy penetrating the atmosphere. Some two million tons of it are now suspended in the air. At the present high rate of increase of air pollution, by A.D. 2040 there will be 228 million tons as a permanent load, which is about as much as was emitted by the explosion of the volcano Tambora, in Sumatra, in 1815. This natural pollution resulted in the coldest weather ever recorded; for two years it caused severe crop shortages and, in some cases, starvation of livestock all over the temperate zone. On the other hand, if particulate emissions are completely controlled and fossil fuels continue to be burned, the carbon dioxide resulting will cause a "greenhouse effect," absorbing the earth's reflected energy and causing a warmer earth. In that event, the Arctic ice cap will melt, the Arctic Ocean will supply much water vapor, and the resultant heavy snows will begin a new ice age. No matter what happens, the earth is going to grow cold, unless we convert to atomic energy.

A contaminant is something foreign that should not be there. A pollutant is a contaminant that does harm to living things. Sometimes a contaminant becomes a pollutant. A tin can by the roadside is an esthetic contaminant, doing no harm but to our sense of beauty; a large pile of tin cans would kill all vegetation and animal life in the soil under it.

There is one simple rule which governs all our thinking about pollutants. Too much is too much. This rule applies to all forms of life. There are three kinds of environmental contaminants. With the first, a little is necessary, a little more is optimal, a lot is toxic. Among them we can mention water, oxygen, nitrogen, air, salt, food, carbon dioxide, the elements necessary for life, both metals and nonmetals, and a variety of organic compounds involved in living processes. All are toxic in large enough amounts. The second type concerns those substances and elements which at ordinary concentrations are not harmful to living things and are not apparently involved in the life processes. Among them are a large number of elements found in

Nature which are relatively inert except in high concentration. The third type involves the toxic elements, trace and bulk, which are toxic in themselves in very small doses and to which living things have low tolerances.

The five toxic trace metals—cadmium, beryllium, antimony, mercury, and lead—are extremely important to the health of the public, being involved in at least half the deaths in the U. S. and much of the disabling diseases. Therefore, they are of public concern—or should be, if we value our lives and our health.

II

METALS AND THE HUMAN BODY

Since his beginning, man has been dependent on metals for life itself. This dependency he shares with all living things on this earth and undoubtedly on other planets.

Man as a functioning structure is composed of twenty-three or so elements. His muscles and organs are made of proteins, which contain carbon, oxygen, and hydrogen, with some nitrogen. His starches and fats consist of carbon, hydrogen, and oxygen. His body is 61 percent oxygen, 23 percent carbon, 10 percent hydrogen, and 2.6 percent nitrogen. Thus, he is nearly three quarters gas.

About 3.4 percent of his weight is made up of eight bulk elements. An average man weighing 70 kilograms (154 lbs) has one kg (2.2 lbs) of calcium, 720 grams (25 ounces) of phosphorus, 140 g each (4.9 oz) of sulfur and potassium, 100 g (3.5 oz) of sodium, 95 g (3.4 oz) of chlorine, 19 g (0.7 oz) of magnesium, and 18 g (0.6 oz) of silicon. Man cannot live without all eight of these elements and all four of the major elements.

The little remaining, less than 0.02 percent (11 g, or about two-fifths of an ounce), is made up of trace elements. At least 24 are found in most of his organs, and 30 or so others are sometimes present in amounts large enough to detect. Eleven of the 54 are necessary for his life or at least for his health; without them he sickens and dies, for they are the very spark plugs of life, the vital reactors which keep him ticking. Eight are metals: vanadium, chromium, manganese, iron, cobalt, copper, zinc, molybdenum. One is a semimetal, selenium; two are nonmetals, iodine and fluorine. Each of these elements performs one or more important functions in the chemistry of life.

Most of the trace metals act as catalysts, causing chemical reactions which would ordinarily not proceed to take place at low temperatures. They are all joined to protein, making enzymes. Magnesium, for example, activates 67 enzymes having to do with energy, as well as many others.

Not all of the essential elements have been identified. There is some speculation that nickel, strontium, and tin may have beneficial effects on mammals. Every decade one or two additional elements are found to be essential. But there is obviously a limit to the number, and we are now approaching it.

Man comes by his heritage of trace elements naturally, for they have been a part of all living things since life began in the primitive seas some two and a half to three billion years ago. The sources of all elements, including the trace metals, are soil, sea, and air.

It is a basic pattern of Nature that plants feed on inorganic matter and animals feed on plants. There are exceptions, of course. Some carnivorous plants feed on insects or even small animals, and many animals feed on other animals which feed on plants. Thus, a carnivorous animal obtains its trace elements from herbivorous animals, who get them from plants.

The earth is about 4.5 billion years old, give or take a couple of hundred million. The ocean, which has been called a dilute solution of practically everything, was formed about 3.3 billion years ago. The first life appeared about 2.7 billion years ago and the first plant life about 2 billion. It is likely that the very first living things took calcium from seawater and combined it with carbon dioxide to make limestone, much as corals do today and as man makes cement and concrete.

The atmosphere at that time was probably mostly nitrogen and methane (marsh gas, which is composed of carbon and hydrogen), with some carbon dioxide, but little oxygen. The absence of oxygen allowed the sun's ultraviolet radiation to penetrate to the surface of the earth, unchecked. Shortwave, ultraviolet radiation breaks down water to oxygen and hydrogen, and carbon dioxide to oxygen and carbon, and it probably started in a small way the entrance of oxygen into the atmosphere, thus beginning the ozone layer. Oxygen combines to form ozone (three oxygen atoms), and today an ozone layer 20 to 40 miles up absorbs most of the lethal, ultraviolet radiation from the sun. Without this barrier no living thing could exist on land or on the surface of the sea.

Of far greater importance to the release of gaseous oxygen were the plants. A billion and a half years ago the warm primeval seas were teeming with plant life, mostly algae. These primitive sea plants had learned to convert carbon dioxide, of which the ocean had enormous stores, to gaseous oxygen and carbon, and to use the carbon in their structures. For this reaction they needed a catalyst and a source of energy. They found magnesium in abundance in the sea, and there was plenty of energy in sunlight. They built up a complex structure, chlorophyll, which is based on magnesium, and which does the same thing today in all green plants—it breaks down carbon dioxide in the air and water to carbon and oxygen.

These small plants which choked the seas changed the atmosphere to what it is today, about 80 percent nitrogen and 20 percent oxygen, with a small amount of carbon dioxide, thus preparing the surface of the earth for life. In addition air contains minute quantities of several other gases and varying amounts of water vapor.

How life began is not known. We can make some informed guesses, however, about its development. The basic structure of living things is protein—large molecules of amino acids, which are made up of carbon, hydrogen, oxygen, and nitrogen. Several amino acids have been made in the laboratory by sparking an atmosphere of methane, carbon dioxide, and ammonia or nitrogen, but they are not alive. Complex proteins have been synthesized—but they are not alive. Simple viruslike particles have been synthesized and exhibit viruslike activities when put inside living cells.

The first amorphous glob of protoplasm which could reproduce itself probably used zinc for growth or protein synthesis, copper for oxidation, iron for oxidation-reduction, and possibly manganese for cell division or reproduction. The first living cell was probably a bacterium, a one-celled organism that does not have a defined nucleus. There are fossil remains of such organisms over two billion years old. To build itself and divide—or procreate—it used the elements available in the primitive seas, especially those with the right properties which were present in reasonable abundance. Besides the "organic" elements making up its protein and fatty structure, it depended on several bulk elements for internal function: potassium, magnesium, calcium, phosphorus—basic stuff for its chemistry. Furthermore, it used and became dependent on several trace elements as catalysts or promoters of basic reactions: zinc for the formation of protein and DNA; iron or copper for oxygen exchanges, or "burning" its food at low temperatures; manganese, for other special purposes.

Through natural selection all life evolves into higher and more complex organisms. To achieve this step up it is necessary to develop special functions different from the mass of life. The tools were there in the sea elements, but they were limited by the number which were abundant enough, and each element was limited by its atomic structure, which could be used only in a few ways.

By the time the first green algae evolved, which took a billion years but changed the face of our planet, additional trace elements were discovered to have uses: boron and, for some species, cobalt, molybdenum, chromium, and vanadium, all present in seawater. These metals provided additional complicated functions for evolution, as Nature experimented.

Sea plants are at the mercy of the seas. They cannot change their environment. It took a billion years or more for the first single-celled animal, which could move, to evolve. That was some eight hundred million years ago. It was a radical change, for now an organism used oxygen and gave off carbon dioxide. The first animal probably was a compromise between plants and animals—an animal with a plant inside. Such creatures exist today in the dinoflagellates, which means "terrible whips." This single-celled animal has a number of whiplike structures which make him swim; in his innards is a chlorophyll-containing simple plant. He is responsible for the "red tide," which blooms, makes our beaches foul, and irritates our eyes and noses. He produces a poison, and is a much more sophisticated animal than his ancestors.

Plants had learned to make structures of calcium, magnesium, and silicon about the time the atmosphere was changed, and animals followed suit. They grew shells, using what was plentiful and would serve: calcium, magnesium, barium, fluorine, silicon, strontium; always experimenting. There was an evolutionary explosion. Plants invaded the land and covered the earth. Where they grew they found the same matter in the soil which they had used in the sea. They took it in through their roots and extracted carbon dioxide from the air. Today most plants require boron, calcium, chlorine, copper, iron, potassium, magnesium, manganese, molybdenum, silicon, and zinc, and some need

cobalt, sodium, selenium, and vanadium, probably chromium, and perhaps iodine. Plants are the only major source of solid elements for
animals living out of water.

It was a great evolutionary advance when animals learned to make
a skeleton. The first skeletons were on the outside of the body, like
those of the modern lobster and shrimp. Skeletons offered protection
and movement, sometimes rapid movement. The energy for this movement was released by the chemical breakdown of phosphate, catalyzed
by magnesium. The skeletons were formed of calcium, silicon, magnesium, phosphorus, and sulfur, with many trace elements incorporated
into them; all were taken from the sea. Insects evolved from skeleton-
ized creatures, and today most insect species, of which there are probably three million, have external skeletons of sorts.

It has been estimated that there are 20 sextillion (20×10^{21}, or 20
followed by 21 zeroes) individual animals and insects living today on
the earth, mostly one-celled animals, and they would weigh, if dried,
4.4 trillion (4.4×10^{12}) pounds or 2 trillion kilograms. They feed on,
and obtain their elements directly or indirectly from, 5 sextillion plants,
mostly single-celled algae, weighing dry 1.1 quadrillion (1.1×10^{15})
kilograms. There are 330,000 named species of plants, and all green
plants convert carbon dioxide into oxygen using sunlight for energy
and magnesium as a catalyst.

Nature is always experimenting and often failing. For example, one
surviving sea animal has a skeleton of strontium sulfate, and another
of barium sulfate (the stuff of barium "meals" and enemas used in
X-raying our innards); these metals worked but didn't catch on generally, for calcium was better. A mollusc has a shell of calcium fluoride,
which worked in this one case.

As we have said, animals use oxygen and give off carbon dioxide,
which is a product of burning sugars and fats. But they must have
some way of fixing oxygen in their blood and tissues, carrying it, and
releasing it where it is needed. There are two metals which will do
that, iron and copper. No other metals have the right properties. The
first chosen by animals was copper, and today most molluscs, insects,
crustacea, and some worms use copper for oxygen exchange. Iron, however, is twice as efficient, efficient enough to allow evolutionary development to mammals.

About half a billion years ago a primitive worm using iron in its
blood copied its many cousins having external skeletons, but instead
built its skeleton as a flexible rod inside its body. At first its nervous

system lay above its skeleton; later it enclosed its nervous system inside its skeleton, for protection. This was a revolutionary idea taking a million years to accomplish, but it made a fast swimmer of this ancestor of ours, especially when tail fins were attached by muscles to the skeleton. Then it began to specialize its nervous system by enlarging the front end to control the tail. Thus was evolved the first prevertebrate with a primitive brain and spinal cord, and the road to man was opened.

So successful was the experiment of enclosing most of the nervous system inside the boney skeleton, and so well protected was the "central nervous system," that the ability of central nerve tissue to heal itself, or regenerate, was lost. A damaged spinal cord never heals and a damaged brain never fully regains its lost function. This is the high price human vertebrates must pay in these days of wars, accidents, and lead poisoning for the development of their brains.

A cousin of our ancestor *Branchiostoma* decided to try vanadium instead of iron to carry oxygen. Vanadium was a poor choice, and the result was the sea squirt! Today, sea squirts have green blood cells containing vanadium, their larvae have a notocord like *Branchiostoma's*, but for a nervous system the adults have only a single ganglion as the first pale ghost of thought.

Branchiostoma probably depended for optional structure and function on 17 elements: four bulk metals—sodium, potassium, magnesium, and calcium; four bulk nonmetals—silicon, phosphorus, sulfur, and chlorine; eight trace metals—vanadium, chromium, molybdenum, manganese, iron, cobalt, copper, and zinc; and one nonmetal—fluorine. Today man and all other mammals depend on the same 17 elements, plus selenium and iodine.

Branchiostoma flourished and evolved, being a fast swimmer and a prolific breeder with few efficient natural enemies. Less than half a billion years ago his descendants decided to invade the swiftly flowing Cambrian rivers, perhaps because of a population explosion, a shortage of food, or a naural urge to explore using their new brain. To go from salt water to fresh was harder than it sounds.

Life began in salinity and has never separated itself from salt water. The oceans increased their saltiness over the eons as huge rivers washed down to the seas enormous quantities of the elements of which the land was made, and living things adapted to an ever saltier environment. Their bodies' interiors had the same composition as seawater, and were in equilibrium with it. To move to fresh water, even brackish water, would be instant death; first they would lose their needed salts, and second, their cells would swell up with water—exactly as an oyster "floated" in fresh water to look plump becomes watery and tasteless—and they would die. They obtained their metals and other elements not only from plants, their food, but also by direct absorption from the sea through their permeable skins.

To move out of the sea, one of these creatures needed, first of all, a skin to exclude fresh water, not absorb it. Because he could not exist in fresh water, he would take salt water with him, inside his skin. A skinful of primitive ocean should work—and did. He had gills—outside lungs—to absorb oxygen from water, salt or fresh, but he could not cover them with skin, and they would absorb fresh water better than salt. So he needed to get rid of this excess water and the water he drank with his food. There was only one way. He evolved a primitive kidney to excrete it. It worked for water, but it was not enough.

Where was he going to get his bulk and trace elements? From food and water, as before, but they were very dilute in fresh water. What little sodium came into his body from fresh water had to be retained, and so later he developed an adrenal gland to do that and a kidney which would excrete only small amounts of salt and other elements. Today the human kidney excretes extra salt eaten, but retains 99.5 percent of the salt that reaches it in blood. The human kidney excretes all the trace metals except cobalt poorly.

Our fish now invaded the rivers, leaving the seas for millions of years to the invertebrates. When they left, the sea was less than a third as salty as it is today, and the salinity of the skinful of ocean they carried with them matched that of the sea. Today, all vertebrates —fish, reptiles, birds, mammals, and man—have the same concentration of salt as was present when the first creatures left the seas.

Fish came back to the oceans while they could still evolve radically. This time they had to meet the problem of too much salt, for the seas were saltier. They invented the chloride gill, which excludes salt, and their kidneys learned to get rid of excesses. But we cannot live on salt water, for our kidneys need lots of water to excrete salt. We have to dilute seawater to one third its strength.

When the fish evolved into amphibians there were few problems, other than breathing air. They could always go back to water to obtain water and a supply of the trace elements that were not adequate in their food. But when they evolved into reptiles, serious problems appeared.

They had to conserve water, often for some time. They had to have available energy stored, and so they stored fat. They had to conserve trace elements in the face of dietary deficiencies and excrete dietary excesses to prevent accumulation. They developed an efficient homeostatic mechanism (*homeo* = same, *static* = state) for each item, for manganese, iron, cobalt, copper, zinc, molybdenum, and fluoride—mechanisms which man and other mammals have today.

Therefore, the same 17 elements, bulk and trace, used by *Branchiostoma* are used by man, and they are vitally important for his life and health. He cannot do without them for long, for he becomes ill when the amounts are marginal and dies when they are very low.

Homeostatic mechanisms involve the kidney, liver, and intestinal tract, and they are exquisitely sensitive. If the body needs an essential element it is absorbed from food; if the body does not need it, it is either rejected by the intestine or excreted in bile or urine. Thus, reasonably elevated environmental exposures in water and food are of no concern to the problem of toxicity from pollution. High exposures, however, can overcome the repulsive mechanisms and lead to accumulation in the body. Too much is too much.

Highly polluted air containing essential trace metals can constitute a hazard since all essential trace elements can be absorbed by the lungs. When a metal is swallowed, the normal homeostatic mechanisms operate. When it is breathed in and absorbed in the body via the lungs, these mechanisms are bypassed. The human kidney excretes metals (except cobalt) poorly. They pile up and theoretically can eventually become toxic, as in the cases of manganese and chromium.

When metals are injected into the body repeatedly, even greater loads are put on the kidneys, which cannot excrete enough to maintain balance. People who are given many blood transfusions may accumulate enough iron to cause disease, for the kidneys can excrete only a milligram or two a day, and a transfusion of 500 milliliters (a bit more than a pint) will contain 80 grams of hemoglobin containing 240 mg of iron. In ordinary amounts iron is not toxic, but repeated transfusions, as for aplastic anemia, can lead to iron overdosage in several years, when the input exceeds the output.

Of greater importance to human health are deficiencies of trace metals and other elements brought about by overrefining and processing food. This is not pollution in the true sense of the word; it is inadvertent overpurification. Too much purity is too much. Deficiencies of metals account for several very common chronic diseases of adults,

such as atherosclerosis (the major cause of death), slow wound healing, occlusive vascular disease, delayed maturity, loss of sense of smell,
malnutrition, and several skin disorders. (These specific diseases will
be discussed in chapter 8.) It is quite likely that some deficiencies
allow other toxic metals to exert their toxicities, thus enhancing hazardous effects.

Homeostatic mechanisms for getting rid of absorbed cobalt, molybdenum, selenium, and fluorine are in the kidney, and toxic effects can
be expected only when the kidneys are poor or the exposure is excessive. There seems to be no way to prevent intestinal absorption of
these elements; they enter the body just as salt and water do. For the
other metals, there is a barrier for iron in the intestine, and for copper,
manganese, and zinc at the intestine and in the liver, preventing the
absorption of excesses. These mechanisms are genetically controlled,
and when the gene is damaged, the mechanism functions poorly. Loss
of the gene for copper control results in "copper storage disease," a
fatal ailment of the liver and nervous system, where copper accumulates to the point of toxicity. Loss of the gene for iron results in "iron
storage disease," a slowly fatal ailment of the liver, pancreas, and skin,
resembling too many blood transfusions. The genetic disease from too
much manganese has not been proven, but it probably lies in the brain.

Aside from these rare hereditary disorders, we have little concern
for the essential elements as hazardous pollutants. Their toxicities are
low and exert themselves only under most unusual conditions. From a
general public health standpoint, they can be largely neglected.

III

METALS AND CIVILIZATION

A side from the vital importance of metals for life, man, the ingenious animal, has been using metals to erect complex civilizations on strong bases of metals. Ever since the first Stone Age man learned to work copper, he has been contaminating himself and his environment with the metals he worked or smoothed. He has been digging them out of deposits and spreading them over the surface of the earth, which he now pollutes at an ever increasing rate with more and more metals.

Each time a new metal or a new alloy was discovered, it introduced a technological revolution. There have been five major ones, based on copper, bronze, iron, steel, and germanium, the last only 20 years ago.

Copper was worked by ancient man some seven or eight thousand years ago. The earliest known artifacts of hammered copper were found in Anatolia, Syria, Iraq, and Iran, dating from the 6th and 5th millennia B.C. Annealing and tooling were developed during the next 500 years, altering the course of human history. After 4000 B.C. melting and casting of copper became common practices in the Near East. Smelting was developed about 3000 B.C.

Copper was, and still is, believed to be endowed with semimagical powers. The alchemist's symbol for copper was the crux ansata, or ankh, ♀, denoting Venus, femaleness, and, to the Egyptians, eternal life (cf. Tutankhamen). In Verdi's opera *Aida* the priests sentence

Rhadames to death by holding the crosses by their rings and pointing them at him, singing in bass unison, "Death! Death! Death!"

In biblical history, the first named metallurgist was Tubal-Cain, of the seventh generation from Adam. "And Zillah [wife of Lameck], she also bare Tubal-Cain, an instructor of every artificer in brass and iron" (Gen. 4:22). This is a mistranslation, for brass was not invented until Roman times, and iron was unknown until about 1000 B.C. One of the instructions for burnt offerings was "But the earthen vessel wherein it is sodden shall be broken: and it be sodden in a brasen pot, it shall be both scoured, and rinsed in water" (Lev. 6:28). The vessel was undoubtedly copper or bronze. Cyprus was the source for Roman and Egyptian copper, and the Latin name *cuprium* is a corruption of *cyprium*. Today some gullible people wear copper bracelets to relieve the pain of arthritis, a survival of the belief in the magical powers of copper. The bracelets are said to be more efficacious if the ore comes from a certain mine!

About 2500 B.C. some unknown genius in Byblos, now Lebanon, melted copper and tin in proportions of 9:1 and changed history. Bronze was strong, tough, and hard, and weapons made of it cut through the softer copper. For the first time a workable hard metal with a low melting point became available to fabricate durable weapons, ornaments, tools, coins, cooking utensils, bells, and statuary.

Bronze is virtually noncorrosive. Millennia later Michelangelo said that clay represents life, plaster death, and bronze immortality. Survival of bronze artifacts for 3,500 years represents a reasonable facsimile of immortality.

Mining and smelting became established and necessary industries, filling the needs of the growing civilization for swords and plowshares, spears and pruning hooks. The costs of transportation necessitated locating these industries at the mines, and adjacent cities grew wealthy. Trade flourished, trading centers expanded, the science of navigation advanced, a capitalistic economy developed, new lands were discovered, and much of Europe and the Near East were opened to commerce. The wealth of the city of Tarshish was founded on Spanish tin and copper. As mines were depleted of ores, exploration for new sources stimulated voyages of discovery, especially for tin, for not many sources are available even today. Trade with Britain, both by sea, a perilous voyage, and by land and Channel, opened up because of Cornish tin (which is still mined), and central Europe was explored because of Moravian tin.

The Age of Bronze lasted well over a thousand years. This multifaceted civilization, which covered the known Mediterranean world, declined with the advent of iron mongering. Bronze was little used in the Western Hemisphere and not at all in Africa, where civilization was retarded for many centuries.

There are 90 naturally occurring elements in the periodic table of the elements, and the universe and its contents are made up of them. Eleven are gases and seven are nonmetals. The rest are metals, 65 of them, and metalloids, seven of them. Not until the beginnings of chemistry in the eighteenth and nineteenth centuries were many of the elements discovered.

The ancients knew of very few. We know they had gold and silver. The Chinese probably had zinc, but the rest of the world didn't until the Romans discovered brass. Zinc was not actually described until 1597.

The ancients knew of antimony and used it for glazing and plating vases, and its black mineral—as kohl—has enhanced the beauty of women's eyes for about 6,000 years. Antimony is toxic to the heart. The Romans made goblets of an antimony-rich alloy and let wine stand in them: the wine absorbed enough antimony to induce vomiting. Tartar emetic is an antimony compound; its use for treating liver flukes has resulted in a high mortality from heart poisoning.

The Romans also knew of arsenic, for it is a by-product of copper, lead, zinc, tin, and gold ores, and they used it medicinally. Many smelterers were probably smitten with arsenic poisoning, for it goes into the air when the ore is heated. Arsenic causes cancer in man.

Lead is of great antiquity; the Romans had lead water pipes which can be seen today in Pompeii, and made amphorae for storing wines and syrups, which dissolved lead. The Roman Empire may have fallen because of chronic lead poisoning from these sources. Only the upper ruling classes could afford such expensive luxuries, and lead took its toll in abortions, stillbirths, and sterility, thus breeding out the aristocracy for 400 years. The artisans were left to do the ruling and fighting, and they were not much good at it. Lead is a low-grade poison, and the fall of Rome probably represents the first example of disaster to a civilization because of pollution. We are faced with the same threat today from lead in gasoline emissions.

In the last decade of the fourteenth century, Chaucer in his *Canterbury Tales* has the Canon's Yeoman speak of alchemy, the state of the art at that time.

I wol yow telle, as was me taught also,
The foure spirites and the bodies sevene,
By ordre, as ofte I herde my lord hem nevene.
The firste spirit quick-silver called is,
The second orpiment,[1] the thridde, y-wis,
Sal armoniak,[2] and the ferthe brimstoon:[3]
The bodies sevene eek, Lo! hem heer anoon:
Sol gold is, and Luna silver we threpe,
Mars yren, Mercurie quick-silver we clepe,
Saturnus leed, and Jupiter is tin,
And Venus coper, by my fader kin!

Today there are 44 so-called industrial metals on which we build our civilization. In Chaucer's time there were nine. Nine were enough for an unmechanized society, but a mechanized one requires metals with a hardness and strength far beyond those provided in pure states. The basic metals of a mechanized economy are copper, for transmitting power, and iron, a fairly soft metal when pure, but one which takes on exceptional strength and hardness when alloyed with arsenic, barium, bismuth, boron, cerium, chromium, cobalt, niobium, manganese, molybdenum, nickel, tellurium, titanium, tungsten, vanadium, and zirconium. Most of these metals were discovered in the eighteenth and nineteenth centuries, although the era of alloying metals had begun before the Bronze Age.

The metals of the ancients are also alloyed to produce substances of special properties. Gold is soft and wears away unless alloyed; so does silver. Mercury makes amalgams with copper, silver, tin, and gold; few people are without such amalgams in their carious teeth, and artificial jewelry is made of them. Lead is alloyed with antimony, tin, copper, and other metals for hardness in shot, bullets, type metal, and water pipes. There is a wide range of copper alloys, which have a great variety of applications.

Whereas the Age of Electricity depends on copper as a conductor, the Electronic Revolution depends on more subtle man-made alloys or compounds and on special properties of a few metals. Germanium crystals are natural rectifiers, and the transistor is made of germanium. Lead selenides and tellurides can also act as semiconductors, and a whole series of new compounds have been made with great difficulty

---

1. Arsenic
2. Ammonium chloride
3. Sulfur

by "solid state" chemists. Who knows when new compounds will be discovered to fill new needs not even imagined today? Or what manifestations of pollution from these compounds will appear?

A Japanese agricultural chemist was called in to discover why the rice in a certain paddy was not heading up. Upstream on the river whose waters irrigated the paddy was a large factory making transistors. He discovered that the factory effluents were dumped into the river, and that the water and the rice stalks contained germanium from the transistor factory. This was a queer effect of germanium. But then he discovered that silicon is essential for plants, and especially for rice plants. Germanium is very like silicon, but different, and it was interfering with the beneficial action of silicon. Problem of pollution solved!

Even some of the newly discovered metals carry mythological or romantic overtones reminiscent of the ancients. Cobalt blue has colored pottery and glass since 1450 B.C., when the Babylonians and the Egyptians used it extensively, and the blue of Ming china and of Venetian glass owes its brilliance to cobalt salts. Cobalt's first contribution to civilization was to beautify. First discovered as a metal in the fifteenth century, cobalt takes its name from the German *Kobold,* meaning hobgoblin, house spirit, or gnome. The skin lesions occurring in miners of arsenical silver-cobalt ores in Saxony and Bohemia—as well as the poisonous arsenical fumes from the smelters—were believed to be caused by these mischievous spirits.

Selenium, discovered in 1817, was named for the Greek *Selene,* or moon. Selene, the moon goddess, was the sister of Helios, the sun, and Eos, the rosy-fingered dawn. According to legend, she was wooed by Pan, and as the price of her favors, she selected a white ram from his flock. She was often shown riding on a ram, a stallion, or a bull. The newborn descendents of these animals, when deficient in selenium, have muscular dystrophy. Selenium has a property which would not surprise the ancient Greeks: it converts light to electricity, and is thus used in the photoelectric cell. The ever-present Xerox machine depends on selenium for copying.

Too much selenium poisons. Certain plants accumulate selenium from soils rich in this element. When cattle or horses eat these plants, they accumulate selenium. Selenium is very like sulfur, but different, and it displaces sulfur in keratins, the substances of hair, nails, and hooves. The resultant selenokeratins are of poor quality; the hair becomes brittle and breaks off, and the hooves rot and literally drop off. This is an example of natural pollution; man has nothing to do with it.

Niobium, a superconductor of electricity and a very hard metal used in alloys, has a history marked by delay, confusion, and neglect. In 1734 John Winthrop (1714–1749), founder of American experimental science and the first scientist to teach at Harvard, sent a sample of black ore discovered near New London to the Royal Society of London. It stayed in the British Museum until 1801, when Charles Hatchett found it, analyzed it, and discovered a new element which he named "Columbium," after Christopher Columbus, or "Columbia." The next year Anders Ekeberg of Sweden isolated tantalum from similar specimens, so named because it was tantalizingly difficult to identify and dissolve. In 1844 Heinrich Rose of Germany separated two metals from the ore columbite, which was believed to have only one; the first he identified as tantalum and the second he named niobium after Niobe, queen of Thebes, daughter of Tantalus, and granddaughter of Zeus.

Niobe was guilty of hubris in referring to her fecundity—she had 12, 14, or 20 children. She boasted of this accomplishment to her great-aunt Leto, who had only one set of twins, Artemis and Apollo, by Zeus, a busy man in this area of divine activity. Jealous Leto, to get back at her, had her two, both expert archers, kill all of Niobe's children—12, 14, or 20—with arrows. Niobe, quite naturally, was grief-stricken, for she was too old to have any more children, and wept so copiously and continuously that Grandfather Zeus immortalized her grief by changing her into a weeping stone—perhaps of columbite. The tale does not say what he did to the jealous aunt or to Artemis, the moon goddess and huntress, and Apollo, the god of music, poetry, prophecy, and medicine, but obviously it was only a mild reprimand, for Artemis later came to be identified with Selene and Apollo with Helios. Incidentally, both niobium and selenium have been found on the moon by the Apollo astronauts. Niobium does not pollute: her tears are pure.

Vanadium was identified in 1831 by Sefstrom and named after Vanadis, the race of Freia, the Norse goddess of beauty. Although many of its compounds are beautifully colored, vanadium workers are likely to get green tongues, hardly a sign of beauty. Vanadium contaminates but does not pollute.

Manganese was used in ancient glass making. It derives its name from the Greek *mangania*, meaning black magic, and *manganon*, machines of war and peace. According to George Cotzias of Brookhaven National Laboratory, who found manganese confusing on close study, its biological activities are "rich in phenomena and lacking in adequate guiding principles," a characterization which smacks of magic. Manganese is seldom a serious pollutant.

Ninety chemical elements occur naturally on the earth. Forty-four industrial metals and semimetals form the base of modern civilization, and some of the 28 other metals and metalloids in the periodic table of the elements are also used industrially. By far the most extensively used is calcium, in cement and concrete. Sodium, in soda lime, salt, and caustic soda has many applications. Potassium has fewer; among other uses it is a constituent of fertilizers. Gold, as a metal, is good for jewelry, the uniforms of "brass hats," gilding the domes of state capitols, plating the nose cones of rockets, and capping teeth. A lot of it is lying useless in bank vaults and Fort Knox, and the gold standard has become a myth (King Midas of Phrygia also collected gold). As yet there is not much use for scandium, yttrium, or hafnium, or for 14 elements formerly named "rare earths," which resemble lanthanum.

Of these 90 elements, nine, of which one is a gas, are naturally radioactive, and five more have some radioactivity built in. (All man-made elements are radioactive; some last only a second or two.) We can measure most of the elements, and if we had instruments sensitive enough we could find all 90 natural ones in the body of man, for they make up the universe and all that is in it.

We can compare the industrial consumption of the essential elements with their amounts in the average, or Reference, Man in order to estimate very roughly potential exposures. In Table III-1 the 16 essential elements are listed in the order of industrial consumption (first column) in thousands of metric tons, followed by the amounts in man's body in milligrams (second column). The concentrations in the earth's crust (third column) and in seawater, where life began (fourth column), are shown in parts per million (ppm) and parts per billion (ppb), respectively. These two columns are an indication of the relative abundances of the essential elements and the ease of obtaining them from soil and sea.

The most popular of the essential metals in our civilization are iron, copper, zinc, manganese, and chromium. There are adequate amounts of them in the environment. As daily intakes the human body needs 200 mg magnesium, 15 mg iron, 15 mg zinc, and 5 mg copper, but only 2–3 mg manganese, 0.5 mg molybdenum, 0.1 mg chromium, 0.1 mg selenium, and 0.04 μg cobalt. Interestingly enough, there are low levels of manganese, chromium, cobalt, selenium, and vanadium in sea-

Table III-1 Elements Essential for Life or Health

Approximate annual U.S. industrial consumption of metals and nonmetals
(1968), their amounts in the human body, and their abundances
on the earth's crust and in seawater.

Element	Industrial consumption (thousands of metric tons)	Amounts in Reference Man (mg)	Igneous rocks, earth's crust (ppm)	Seawater (ppb)	Human disease from excess
Iron	109,000	4,200	56,300	10	Iron-storage disease (genetic)
Calcium	86,273	1,000,000	41,500	400,000	
Sodium	15,091	100,000	28,600	10,500,000	
Potassium	3,230	140,000	20,900	380,000	
Copper	1,400	72	55	3	Copper-storage disease (genetic)
Zinc	1,278	2,300	70	10	
Manganese	1,050	12	1,000	2	Manganism, miners
Fluorine[1]	587	2,600	700	1,300	Fluorosis
Chromium	459	1.5	100	0.5	Cancer (chromates), workers
Nickel[2]	170	10	75	5.4	Nickel carbonyl cancer, workers
Magnesium	89	19,000	23,300	1,350,000	
Molybdenum	25	9	1.5	10	
Cobalt	6	1.5	25	0.27	
Strontium[3]	6	320	375	8,100	
Vanadium	5	18	135	2	
Iodine	—	11	0.5	60	
Selenium	0.5	13	0.05	0.09	Cancer, rats

1. Nonmetal, essential for healthy bones and teeth.
2. Essential for birds, possibly essential for mammals.
3. Possibly essential.
SOURCES: Bowen 1966, Tipton (in press), and *Mineral Facts and Problems*.

water; their relatively low abundances in the environment of developing living things probably limited their usefulness to a few functions. It is also interesting that the greatest toxicities are shown by the elements of lowest abundance on the earth's crust and in seawater: selenium, molybdenum, iodine, and cobalt, although their toxicities are very low compared to those of some other elements.

The table also lists the major human diseases resulting from too much of the essential elements, two caused by gene deficiencies (iron and copper), and one caused by excessive exposure to dust in mines (manganese). It lists three that cause cancer, two occurring in man (lung cancer from dusts containing chromates, and lung and nasal cancer from nickel carbonyl).

To those who are concerned that fluoridating water supplies with one ppm fluorine as sodium fluoride may be a form of pollution dangerous to people's health, reassurance to the contrary may be found in the table. Fluorine is very abundant on the earth's crust at 700 ppm, and at 1.3 ppm is the most abundant trace element in seawater after strontium. It is obvious that no toxicity has been exerted on land or sea except by drinking water with much larger concentrations, and that 1.0 ppm fluoride in water is perfectly safe.

Aside from exposures of workers directly handling metals and their compounds, the general population comes into contact with many potentially harmful metals in one way or another. There are some 35 or 40 metals in the air, most coming from the burning of coal and oil. Rubber contains antimony, cadmium, lead, selenium, tin, and barium—an imposing array of additives. Ceramics and pottery glazes have antimony, beryllium, lead, barium, nickel, and zirconium. Paints and pigments contain antimony, cadmium, lead, mercury, selenium, barium, cobalt, vanadium, zinc, manganese, tungsten, and zirconium, to name a few. One cannot live in the world today without frequent continuous contact with many metals and metalloids. We get them in air, food, and drink, in small amounts, to be sure, but the amounts can get fairly large if they accumulate in our bodies over a lifetime.

Our problem is to decide which ones are harmless and which ones are harmful. To decide, we must examine each one separately, for each element is a unique atom like no other, and each behaves in a way different from all others. The problem is serious when it concerns our health, for chronic ill health is one of man's heaviest crosses, and not many people live to their allotted life span of 90–100 years without disease.

Table III-2. Toxicity of Important Elements

Approximate annual U.S. industrial consumption of metals and nonmetals
(1968), their amounts in the human body, and their abundances
on the earth's crust and in seawater.

Element	Industrial consumption (thousands of metric tons)	Amounts in Reference Man (mg)[1]	Igneous rocks, earth's crust (ppm)	Seawater (ppb)	Human disease from excess
Toxic to Living Things					
Lead	816	*121(9–480)*	10	0.03	Plumbism
Antimony	19	*8*	0.2	0.33	Heart Disease
Beryllium	0.3	0.04	2.8	0.0006	Beryllosis
Cadmium	7	*38*	0.2	0.11	Hypertension, emphysema
Mercury	3	13	0.1	0.03	Poisoning
Slightly Toxic to Some Life Processes					
Tin	59	6	2	3	
Arsenic	22	*18*	1.8	3	Cancer
Tungsten	7	+	1.5	0.1	
Germanium	11	+	5.4	0.07	
Uranium	2.7	0.09	2.7	3	Kidney disease, animals
Bismuth	1	0.2	0.2	0.017	
Tellurium	0.1	*8*	0.001	—	
Palladium	0.02	+	0.01	—	
Rhodium	0.002	+	0.001	—	
Probably Inert to Living Things					
Aluminum	3,534	61	82,300	10	
Barium	700	22	425	30	Baritosis
Titanium	413	9	5,700	1	
Zirconium	61	420	165	0.022	
Lithium	2.6	2	20	180	
Silver	0.03	0.8	0.1	0.3	Argyria
Niobium	2	110	20	0.01	
Boron[2]	70	14	10	4,600	

1. Numbers in italics are considered to be larger than normal for uncontaminated man.
2. Essential for plants.
SOURCES: Bowen 1966, Schroeder 1965, Tipton (in press), and *Mineral Facts and Problems*.

Many studies of toxicity of elements have been made. The most pertinent to the problem of the effects of pollution are those in which animals are exposed to low doses daily throughout their entire lives. The observed results are compared to those for unexposed, control animals. To conduct such studies we need a metal-free environment, which must be made of wood. Under these conditions, some 3,500 mice have been exposed to low doses of each of 30 metals and elements for life, up to three years, and some 2,000 rats to 16 elements for life, up to four years or more. With older experiments and this experience, we can classify important elements according to severe toxicity, slight toxicity, and no toxicity (Table III-2).

The table clearly shows that the five toxic elements are very scarce, and that the nine slightly toxic ones occur in low concentrations on the earth's crust and in seawater. Therefore, developing life forms never learned to cope with these 14 elements, for there was no need. Because of human profligacy, there is more lead, antimony, cadmium, mercury, arsenic, and tellurium in human tissues than there probably would be in uncontaminated man, and some of these elements constitute serious hazards to human health, especially cadmium and lead, probably mercury, antimony, and beryllium, and possibly germanium.

Of the eight inert metals in Table III-2, all but silver are quite abundant on the earth's crust, and four are abundant in seawater. Problems of pollution are minimal. Aluminum, barium, titanium, and zirconium are natural pollutants from dust and coal, like iron, calcium, magnesium, and strontium, and man has been breathing them since the beginning. They probably do no harm.

Lifetime studies of small doses have not been done for the industrial metals bismuth, cerium, cesium, lanthanum, lithium, osmium, rhenium, rubidium, ruthenium, tantalum, and thallium (very toxic). Their concentrations in human tissues are too low to be deemed toxic or causal of disease at this time. Perhaps with newer uses of these metals such experiments will become mandatory. They are now needed for ruthenium, which has new applications in the automobile industry.

Table III-2 also shows those metals which are present in the human body in larger amounts than are justified by natural environmental exposures, and are therefore candidates for pollutants. Lead and cadmium are the worst offenders; antimony, arsenic, and tellurium are also present in abnormally large amounts in relation to their crustal concentrations. These five elements, plus beryllium and nickel, are those on which we must focus our attention.

IV

ENVIRONMENTAL CONTAMINATION

BY NATURE AND INDUSTRY

Man has contaminated his environment with metals and other elements ever since he discovered fire. With ever increasing numbers of fires burning coal and oil, smoke emissions into the air have become the greatest source of contamination, and the contents of smoke the major pollutants. What goes up must come down, to be breathed in air, to be drunk in water, and to be eaten in vegetable foods. There are special instances of pollution by single metals, as increments to the burning of fossil fuels, but smoke ranks the largest in general terms.

Nature can pollute the air with fire. The explosion of the volcano Tambora in 1815 put 220 million tons of particulates into the stratosphere, which so chilled the world that the next two years were the coldest in history, as stated in chapter 1. Man is now emitting 2 million tons but at the present rate of increase of consumption of fossil fuels, he will emit 228 million tons into the stratosphere as a permanent load within the next 65 years, and population control through starvation will be a reality. The lowered temperature will result in vast losses of crops, as it did in 1816, losses of cattle and sheep because of a lack of forage, and a worldwide food crisis in the temperate zones. Another volcano would do it now. Likewise forest fires lit by lightning can emit much smoke.

Aside from periodic explosions of major volcanoes, which occur on the average of five per century (Krakatau in 1883 emitted 50 million tons), Nature is constantly eroding the land and contaminating the waters of the rivers and lakes with metals and other elements. The extent of this natural "pollution" (for it probably does little harm) is enormous. Added to what man has done, though, it may make the difference between harm and harmlessness in some cases.

As we are concerned with the toxicity of elements resulting from man's activities, let us compare what man has done and what Nature is doing to contaminate our environment. To do so, we will divide the metals and a few trace elements into two groups, those necessary for life and those not necessary and possibly toxic.

Everything is toxic in excessive amounts. Compared with usual daily amounts eaten or drunk, water is the most toxic of all, for if we were to drink four times our usual intake of three quarts a day, we would die of uremia in a few days. We can take with impunity four times our daily amounts of iron, zinc, manganese, chromium, copper, molybdenum, iodine, and fluorine, although we might have some trouble with sodium.

Nature "contaminates" the air with enormous quantities of those metals which are abundant on the surface of the earth. Winds account for most of the aluminum, barium, strontium, silicon, titanium, and much of the iron, manganese, and chromium in the air. These metals are all abundant on the face of the earth, and dusts contain them. These dusts settle to the ground whence they came. Man breathes these dusts, and they collect in his lungs, and always have, long before the words "air pollution" were thought of. At ordinary exposures they do little harm, if any.

Nature also "contaminates" the oceans with vast amounts of elements from the land, by dissolving them in rainwater and pouring them in rivers to the seas, or by carrying down in the rivers suspended solids of earth. Some 334 trillion (334,000,000,000,000) tons of water evaporate from the ocean each year, of which 100 trillion tons fall on land as rain. More than half evaporates off again, leaving the rivers to carry 37 trillion tons of fresh water containing 4.5 billion tons of sediments to the sea. Thus are deltas formed. Fresh water also carries large amounts of dissolved elements, which have been making the sea saltier.

Fortunately for life in the ocean, salt water is a most efficient scavenger of metals. We must go to Table IV-1 to illustrate this point.

Of the 16 essential bulk and trace metals measured, only sodium, magnesium, potassium, and boron are retained in seawater to any sizeable extent (1.6 to 46 percent), thus enriching the seas.

Table IV-1. Essential Elements in Seawater and in Air

Amounts added by rivers to the oceans every year by weathering, amounts which stay in the ocean in solution, and amounts emitted to the air from the burning of fossil fuels.

	Nature			Retained in seawater		Man's activities		
	Rivers	Sedi- ments	Total	(%)	(metric tons)	Coal	Oil	Total
	(thousands of metric tons)					(thousands of metric tons)		
Sodium	230,000	57,000	287,000	40.0	114,800*	280	0.33	280
Magnesium	148,000	42,000	190,000	5.1	9,690*	280	0.02	280
Potassium	83,000	48,000	131,000	1.6	2,096*	140	—	140
Calcium	540,000	70,000	610,000	0.9	5,490*	1,400	0.82	1,400
Vanadium	32	280	312	0.001	3.1	3.5	8.2	12
Chromium	36	200	236	0.00004	0.9	1.4	0.05	1.5
Manganese	250	2,000	2,250	0.0002	4.5	7	0.02	7
Iron	24,000	100,000	124,000	0.00002	24.8	1,400	0.41	1,400
Cobalt	7.2	8	15	0.001	0.15	0.7	0.03	0.7
Nickel	11	160	171	0.005	8.5	2.1	1.6	3.7
Copper	250	80	330	0.005	16.5	2.1	0.023	2.1
Zinc	720	80	800	0.01	80	7	0.04	7
Boron	360	—	360	46.0	165*	10.5	0.0003	10.5
Fluorine	3,300	—	3,300	0.2	6,600	10.8	—	10.8
Selenium	0.14	10	10	0.16	16	0.42	0.03	0.45
Molybdenum	36	28	64	0.6	384	0.7	1.6	2.3

* Thousands of metric tons.
SOURCE: Calculated from Bertine and Goldberg 1971.

In the first three columns are the amounts in thousands of metric tons of essential elements that have been weathered and are being carried downstream to the seas by the rivers, in solution or as suspended sediments. The largest amounts are for calcium, sodium, magnesium, potassium, iron, fluorine, and manganese. The next two columns show the percent of the element and the tonnage that is retained in solution in seawater.

The rest goes to the bottom, with only very small amounts dissolved in seawater. For example, 124 million tons of iron are carried to the sea by rivers every year; about 25 tons stay in seawater. As there are 1.42 quintillion (1.42×10^{18}) tons of ocean water, this addition is negligible. So is the addition of fluorine, which is quite soluble in fresh water; 3.3 million tons go in and 6,600 tons stay dissolved, a drop in the ocean bucket. Every million years the ocean gets 0.0081 percent saltier, from sodium, under present rainfall conditions. Half a billion years ago much more sodium was added to the sea as the lands eroded under torrential rains.

How does this record compare with man's activities in burning fossil fuels and spewing them into the air?

When we look at the essential elements, man's efforts are puny compared to Nature's. What goes up in smoke must come down and eventually get into the ocean. The last three columns of the table show that man contaminates the air with only about a hundredth as much iron as Nature carries to the ocean, although the total tonnage in air is much greater than what stays in solution in the sea. Only in the cases of vanadium, cobalt, selenium, and molybdenum does man put into the air more than three percent of what Nature puts into the ocean (3.6%–4.7%).

We have not discussed the toxic elements, which are our prime concern. It follows from the evolution of living things that elements abundant in seawater and on land are not toxic, especially those used by living things. On the other hand, elements of low abundance in seawater and on land may or may not be toxic depending on how soluble they are, for until now life has been in only limited contact with them.

When the molten earth solidified, released its water, and began to weather, many metals and other elements were deposited in insoluble forms in cracks and seams, from which they are now mined. In their natural deposits, they were effectively removed from contact with life, and when they did enter the ocean, they were quickly scavenged and deposited on the bottom. Small plants and animals did, and do, most of the scavenging. Thus the earth and the seas were relatively depleted of these elements. But man has mined these deposits and spread them over the surface of the earth in ever larger amounts.

Now let us see how man contaminates his air with these less abundant elements. All of them are found in coal, the product of thousands of years of massive vegetation, and in oil, the product of animal life. In Table IV-2 we see that man's emissions are not all puny compared

Table IV-2. Toxic and Inert Elements in Seawater and in Air

Amounts added by rivers to the oceans every year by weathering, amounts which stay in the ocean water, and amounts emitted to the air from the burning of fossil fuels.

	Nature			Retained in seawater		Man's activities		
	Rivers	Sedi-ments	Total	(%)	(metric tons)	Coal	Oil	Total
	(thousands of metric tons)					(thousands of metric tons)		
Toxic								
Cadmium	300	0.5	300	0.05	150	0.04	0.002	0.042
Beryllium	4	5.6	9.6	0.00002	0.002	0.41	0.0003	0.41
Antimony	0.13	—	0.13	0.15	0.195	0.25	—	0.25
Mercury	2.5	1.0	3.5	0.03	1.05	10	0.6	10.6
Lead	110	21	131	0.000002	0.003	3.5[1]	400[2]	408
Slightly Toxic								
Tin	—	11	11	0.1	11	0.28	0.002	0.28
Barium	360	500	860	0.006	51.6	70	0.02	70
Germanium	—	12	12	0.004	0.5	0.7	0.0002	0.7
Gallium	3	30	33	0.0002	0.07	1.0	0.002	1.0
Arsenic	72	—	72	0.15	108	0.7	0.002	0.7
Yttrium	25	60	85	0.0008	0.7	1.4	0.0002	1.4
Lanthanum	7.2	40	47	0.0009	0.4	1.4	0.0008	1.4
Bismuth	—	0.6	0.6	0.01	0.06	0.75	—	0.75
Scandium	0.14	10	10	<0.00002	<0.002	0.7	0.0002	0.7
Nontoxic								
Aluminum	14,000	140,000	154,000	0.00001	15.4	1,400	0.08	1,400
Titanium	108	9,000	9,100	0.00002	1.8	70	0.02	70
Strontium	1,800	600	2,400	1.9	45,600	70	0.02	70

1. Plus 5,190 tons from smelting, refining, and incineration.
2. 400,000 tons from gasoline plus 50 tons from petroleum.
SOURCES: Calculated from Bertine and Goldberg 1971, and Bowen 1966.

to Nature's effluents. More lead is emitted into the air, mainly from gasoline additives, than Nature carries to the seas. Lead is an outstanding pollutant. About three times as much mercury enters the air as the waters dissolve. We put less than one percent each of Nature's effluents of cadmium, aluminum, titanium, and arsenic into the air, but we add more antimony and bismuth, 8 percent as much barium, 7 percent as much scandium, 6 percent as much germanium, and 4 percent as much beryllium as Nature dissolves from the land. Most of these metals come from coal. What we are doing is emitting into the air relatively little of the nontoxic trace elements and emitting relatively more of some but not all of the toxic ones, thus bending the balance of Nature. Other sources of exposure make the problem much more serious.

The oceans are not polluted, despite man's activities. The table shows that of the elements entering the sea every year from Nature's weathering, about 108 tons of arsenic, 150 tons of cadmium, 52 tons of barium, a ton of mercury, and 7 lbs of lead stay dissolved. If all the airborne metals coming from fossil fuels were added to seawater (which is what eventually happens), an additional 42 tons of cadmium, 410 tons of beryllium, 250 tons of antimony, 10,600 tons of mercury, and more than 200,000 tons of lead would enter the oceans, of which all but small percentages would be precipitated to the bottom. These toxic metals could be picked up by single-celled plants and animals and enter the food chain of fish, but only lead and mercury would represent amounts significantly larger than Nature's efforts.

The burning of fossil fuels is not the only method of contaminating the modern environment, but it is most difficult to measure the total amounts of metals dissolved in rivers from mine tailings, industrial wastes, and garbage, or of airborne metals from incinerators and other sources. We can go to extremes, however, and see what would happen to the oceans if the total amount of all metals mined every year were dissolved in fresh water and added to the seas (Table IV-3). Not much.

The first column shows an estimate of the amount of metal mined per year, as of 1965. The second column shows the concentrations which would be found in fresh waters if all the metal mined were dissolved and dumped down the drain to enter the rivers and lakes of the earth. The concentrations of lead, chromium, and copper (here expressed in parts per trillion) would be suspiciously elevated but

Table IV-3. Potential Pollution of Fresh Water and Seawater

Amounts of metals mined each year and their concentrations in fresh water
and seawater if all the world's yearly mining and particulates
from burning were dissolved and dumped down the drain.

Metal	Total mined per year	Fresh water	Dissolved in seawater	From air, dissolved in seawater
	(thousands of metric tons)	(parts per trillion)	(metric tons)	(metric tons)
Cadmium	10	20	5	0.021
Mercury	10	20	3	3.18
Antimony	45	100	67.5	0.4
Lead	2,200	4,000	0.044	0.004
Tin	170	300	170	0.28
Thallium	0.03	0.06	0.0006	—
Bismuth	1.4	3	0.14	0.075
Germanium	0.08	0.18	0.003	0.03
Nickel	300	600	15	0.38
Tungsten	30	60	1.8	—
Chromium	2,000	4,000	0.8	0.0006
Copper	4,000	8,000	200	0.1
Arsenic	2.7	5	4.05	1.05

Total mass of fresh water = 510 trillion (510×10^{12}) tons
Total mass of seawater = 1.42 quintillion (1.42×10^{18}) tons
SOURCE: Computed from Bowen 1966 and Tables IV-1 and IV-2 above.

would still be far below official limits for drinking water (which are
expressed in parts per billion). The third column shows the tonnage
which would remain dissolved in the oceans after the rivers had
flowed into the seas. The amounts remaining in seawater are negligible; the difference between column one and column three indicates
the enormous ability of seawater to scavenge and precipitate metals
to the bottom. The fourth column shows the amounts that would remain dissolved in seawater after the rains had brought down the
particulates of metals in the air from the burning of fossil fuels.

Even when we add the potential pollution of metals from burning, as it eventually reaches the ocean, the overall picture is little changed. Less than a ton of most metals will be added to those dissolved in seawater. Therefore, we can say categorically that the vast oceans are not polluted, or even contaminated, by metals resulting from man's industrial activities. Nor would they be polluted in a thousand years at the present rate of activity. This bugaboo should be laid to rest. In a thousand years, the concentration of mercury in the seas would reach 4.35 parts per quadrillion added to the present level of 30 parts per trillion, or 30,004.35 parts per quadrillion.

By no means do I mean to imply that local areas are not polluted. There are hundreds of examples of serious metallic pollution of stretches of rivers, lakes, bays, estuaries, sewage treatment plants, air, and soil. They do serious harm to plants and animals in these areas, and to man. But the overall picture, while potentially hazardous, is generally not as bad as depicted by some alarmists. For example, samples of water from 720 major lakes and rivers in the United States were analyzed (see Table VII-1). Thirty-one bodies of water contained more cadmium than is considered allowable for drinking water, nine had more arsenic, two had more lead, and one was close to the allowable limit for lead. In none was there more mercury. Once detected, those waters can be cleaned up. Some municipal waters have concentrations of metals exceeding allowable limits, but there are relatively few. They can be cleaned up.

The problem of pollution, while basically a matter of carelessness, neglect, lack of knowledge, or callous indifference to others, is not one large problem but a number of smaller problems. It can be solved only by treating each pollutant separately. The source, the medium of emission to the environment, the distribution, and the entrance into the body of man must be considered for each. Above all, the health hazards for human beings are the most important aspect of the problem.

Thus we see that there are several ways by which the environment becomes contaminated with metals and other elements. Nature causes the largest amount of contamination of water with most metals, but they are in the main either beneficial to life or harmless, and most of them are scavenged by the ocean. Nature contaminates the air mainly through volcanic eruptions, which can seriously pollute the environment, and to a slight extent through forest fires. Volcanoes can release toxic metals from the earth's (molten) magma, but wood fires do not emit more than traces of toxic metals, for the wood would not have grown had there been much.

Among man's pollutants, coal also contains relatively small concentrations of toxic metals for the same reason. It has a good bit of the metals of relatively low toxicity, such as lead and mercury, but not much cadmium, beryllium, antimony, or arsenic. Except for lead, which is added in large amounts to gasoline, crude petroleum products contain little toxic metal and their distillates almost none. If the living organisms which synthesized petroleum had accumulated toxic metals from the environment, they would have died. Of the toxic metals, the largest contaminants from coal and oil are mercury, lead, and barium.

Contamination of air with specific metals in local areas comes from refineries emitting smokes and gases. The white plume going miles into the air above a zinc refinery contains, besides zinc, cadmium and lead in large amounts. Cadmium and lead can also come from copper refineries. Water is always contaminated by mine tailings draining into rivers; they can be anything which is mined and is soluble in water. There are no good estimates available on the total amounts of the various metals which contaminate air and water directly, although part can be estimated from washings of electroplated metal parts; such washings are poured down the drain.

Each pollutant has a different pathway, and several pathways may join and then diverge. As an example, if people litter their streets with paper and cartons, there is little hazard to health, although esthetically the litter is revolting. When they also litter with garbage, rats breed, and rats are a hazard to health because they carry infectious diseases. When automobile and truck exhausts fill the dirty streets with lead, and children playing inhale and swallow the lead, the stage is set for lead poisoning. When the same children eat flaking lead paint from interior surfaces—so painted by unknowing people— they become poisoned. Lead-poisoned children may become mentally retarded, requiring care for the rest of their lives. In this case the blame is laid directly at the feet of the painters and contractors and indirectly at the petroleum industry for putting lead in gasoline. The painters may have been unwitting, but the gasoline industry is callous.

In only two cases is the amount of an element emitted into air from the burning of fossil fuels greater than the *potential* exposure from dumping down the drain all of that element mined annually. There is more germanium and mercury contaminating the oceans from air than from water. In all other cases, mining contributes many times more metal than does the burning of coal and oil.

These considerations are the background of our discussions of each metal, its effect on man, and possible methods for eliminating hazardous metals.

V

LEAD, THE LEAD INDUSTRY, AND

MAN'S HEALTH

Lead is a protoplasmic poison. One must have a nineteenth-century faith in the virtue of business to believe that any large industry puts the welfare and health of the people ahead of its profits and sales. The lead industry is no exception. But let us examine it as though it were virtuous.

The December 1971 issue of *Natural History* printed the following exchange in its "Letters" column:*

WHICH WAY WILL YOU BE LEAD?

What is a reader to believe?
I received in today's mail a clipping from the September 19 *Morristown* (Tennessee) *Citizen Tribune* and have just finished reading:

The current move toward unleaded gasoline for cars is the kind of misleading and irrelevant course of action that makes a solution to air pollution seem impossible. First of all, there is no evidence that lead in the atmosphere, from automobiles or any other source, poses a health hazard. Spokesmen for the U.S. Surgeon General, the American Medical Association, the U.S. Public Health Service, and the World

---

* Copyright © The American Museum of Natural History, 1971.

Health Organization have all said that lead in the atmosphere is not now nor foreseen to be a threat to health.

Then the October, 1971, issue of Natural History Magazine arrived. And in the article "Greetings from Los Angeles," by Ira J. Winn, I read:

Lead is also a deadly poison. . . . Its effects are cumulative and result in a large number of health problems, beginning with such symptoms as fatigue, dizziness, cramps, and headaches, and eventually leading to a variety of disorders that can end in paralysis, brain damage, and death. A number of scientists suspect low-level lead poisoning as a cause of many of our ills.

Confusing at least, wouldn't you say?

James L. Bailey
Director, Educational Service
Tennessee Department of Conservation

Editor's Note:

Mr. Bailey's letter raises some interesting questions. To satisfy our curiosity, we tracked down the story that appeared in the Morristown newspaper. We found it had come from *Editor's Digest* in New York, a division of Planned Communication Services, Inc., a company that writes and distributes stories to small-circulation newspapers on behalf of corporate and industrial clients. This story, it turns out, originated with the Lead Industries Association (LIA) in New York. LIA's public relations firm, Hill & Knowlton, funneled the information for the story to *Editor's Digest.* We note in passing that readers of the Morristown paper had no way of knowing that a story about lead in air pollution originated with the Lead Industries Association.

The Author Replies:

The campaign against environmentalism is now moving into high gear and readers should be advised to expect an outpouring of half-truths, invective, obfuscation, and plain old lies as the polluters go all out to convince the people that all is well with things just as they are; that anyone who criticizes the system is a disaster-monger; and that industry is prepared to lead the way in the antipollution struggle and doesn't need any advice from onlookers or any regulation from the federal government.

But the carefully planted and phony stories aimed at lulling growing public fear of airborne lead poisoning from automobile exhaust are a superb example of just why the general public and the scientific community should have a deciding voice in matters of environmental concern and why federal control over air polluters is essential if we are to avoid a public health disaster.

These planted fables show how the polluters' campaign gears up when the economic chips are down. Let's look at how the deck is really stacked, as seen by some genuine experts.

Dr. Henry A. Schroeder of the Trace Element Laboratory of
Dartmouth College, one of the world's foremost authorities on the effects
of metals on the human body, states flatly:

> Airborne lead is largely a result of anti-knock additives to gasoline
> in the form of tetraethyl or tetramethyl lead. . . . Lead offers a
> potential and imminent hazard to human health. . . . We have found
> enough lead in vegetation growing beside a secondary highway (up
> to 200 parts per million) to abort a cow subsisting on this vegetation.
> The concentration has trebled in six years. . . . Evidence of a
> biochemical abnormality in persons exposed to urban air concentrations
> of lead is beginning to appear. There is little doubt that at the present
> rate of pollution, diseases due to lead toxicity will emerge within
> a few years.

R. A. Kehoe, a specialist in lead toxicology, reports that 30 to 40 percent of lead inhaled by humans is absorbed into the body, while only 5
percent of the lead intake from water and food is absorbed.

Dr. John R. Goldsmith, head of epidemiology for the California
Department of Public Health, finds that the blood lead levels in persons
living near a freeway are markedly higher than in persons living in
relatively clean coastal air a mile from that freeway.

Tsaihwa J. Chow, a specialist on lead at the Scripps Institution of
Oceanography, reports that concentrations of lead in plants growing near
highways are many times higher than what is found away from traffic.
Dr. Chow, an internationally acknowledged expert on lead and a
consultant to the National Research Council, has taken strong public
exception to a news release issued by the Council about its 500-page lead
study published last summer. Chow, who had worked on the study,
charged that the release distorted the essence of the report and gave "a
false sense of security and well-being." Dr. Chow made clear that airborne
lead threatens health in most urban areas, including virtually the entire
Los Angeles basin, and the hazard is not limited to central cities. No
other toxic chemical pollutant has accumulated in man to the extent of
lead, which poses "an identifiable, current threat to the general
population," he said.

Airborne lead in the cities studied, Chow concludes, has been
increasing about 5 percent per year or 55.7 percent for Los Angeles in
the 1960s alone.

Peculiarly, the NRC press release indicated that "some" cities studied
showed increases, whereas the report itself states that 17 out of the 19
sites studied showed large increases in airborne lead. The news release
stated that the average American ingests more lead than he inhales, a
statement sure to mislead the public simply because most lead taken
through the mouth is excreted, while that taken into the lungs is retained
to a much higher degree.

The release argued that two to three times as much lead is added to the environment in the form of paint pigments and metal products as in the form of gasoline additives, a fact of dubious value considering that relatively few people munch on paint or eat car batteries. And to top off the simplistic nonsense that was handed out to the press and widely circulated, the NRC release—in part written by a former employee of one of the world's leading producers of lead additives for gasoline—noted that "only young children and certain groups of workers face potential health hazards from airborne lead." Perhaps to the polluters and profiteers, *only* is a nice clean adjective to place in front of lead-poisoned children.

In sum, I can only remind the reading public that the price of clean air, like the price of liberty, is eternal vigilance. The foxes are still trying to guard the henhouse, and if you don't believe me, go outside one stinking day in Los Angeles, take a deep breath, and remember what Admiral Rickover observed: "You don't have to be a hen to smell a rotten egg."

Ira J. Winn

So much for virtue.

What Mr. Winn did not say, but could have, is that in October 1970 the Secretary of HEW, Robert Finch, issued a ban on leaded gasoline in government automobiles as a result of a petition from the Environmental Defense Fund, that the American Medical Association has made no official statement on the health hazards of airborne lead, and that the World Health Organization is now studying the problem. Some of the preliminary conclusions reached by WHO's Working Commission on Health Hazards of Toxic Substances in Water are:

In most surveys on the general population, a small proportion of the persons studied show lead-in-blood values high enough for some biological response(s) to be expected. The reason why these values are so high is usually not explained: however some additional (occupational or otherwise abnormal) exposure is the most probable reason. Be that as it may, a cross section of the general (usually urban) population as a rule thus reveals individuals exposed to potentially harmful lead doses.

It is obvious that recommendations for a maximal daily intake are needed. Such recommendations should contain a safety margin broad enough to protect even the most vulnerable part of the population. But recommendations are of little use if there is no practical possibility to follow them. Thus the most important task is to decrease or eliminate pollution at the source. This is a tremendous task and requires international agreements and collaboration immediately since the persistent nature of the pollutant makes it difficult to cure damage already done.

For nearly 40 years the definitive biological work on lead and the tolerable doses for man were the subject of study by the Kettering Institute in Cincinnati. There, rules and regulations on exposure and handling of tetraethyl lead and its relatives were carefully made, to protect workers handling this most insidious toxin, which is added to gasoline. I wanted half an ounce for some student experiments: It had to be brought in a car from duPont in Wilmington, Delaware, to Hanover, New Hampshire, some 400 miles, for it could not be shipped. It arrived in a quart-sized pressure bottle, like a fire extinguisher, with a large brass nut on the top, surrounded by compressed insulation in a sealed gallon can, surrounded by much insulation in a carefully built thick wooden box screwed together. We used about a tenth of it. What to do with the rest? According to instructions, it could not be poured down the drain or on the ground—too poisonous. It could not be buried—the bottle might corrode and leak. Bottom of the river—oh, no. The only "safe" thing we could do was to pour it into the gas tank, where 270,000 tons of lead in tetraethyl lead were poured in the United States last year.

The Kettering Laboratory also worked out the amounts of lead in food and water which could be tolerated by man without getting lead poisoning and without accumulating in his body. At 0.6 milligrams a day, a state of balance was said to have been achieved; what went in came out. It was concluded that lead was no problem: the average intake was about 0.3 mg and only 5 to 10 percent of what is eaten is absorbed into the body. What was partly neglected were the large amounts in air from motor vehicle exhausts, much of which are absorbed—perhaps half of what one breathes. The Kettering Laboratory was largely supported by Ethyl Corp., which, with duPont, processes the lead we breathe. For 40 years everything went smoothly; lead in air was breathed blithely, and lead was innocuous.

In 1965 Clair C. Patterson of Caltech published a long and carefully done article on lead in the environment. He said that the body burden of lead was 200 times what it should be, and that lead from gasoline was everywhere, on the peaks of the Sierras and in the depths of the Pacific. He was soon to go to Greenland to measure the sharp accumulation of lead in glacier ice laid down since 1940, which was due to airborne lead.

A meeting was called in Washington to consider atmospheric lead. It was heavily overloaded with representatives of industry, lead, and petroleum, and the few scientists concerned about the problem were talked down and outvoted. The press, however, asked some embarrassing questions, thinking it curious that a laboratory supported by industry should be the arbiter and monitor of what it sold. Patterson was not invited. Dr. Harriet L. Hardy, who knows more about lead than anyone, was the principal gadfly, but she was talked down or disregarded.

Industry is so virtuous.

Isabel H. Tipton had given me raw data on lead in bones and tissues of some 400 subjects from here and around the world, some of which I had collected as part of a much larger project in 1957. We put them together and found that lead at exposures existing in the United States from 1951 to 1957 accumulated in human tissues with age, whereas at exposures in 17 other countries it did not. These results were contrary to what the Kettering Laboratory claimed, i.e., no accumulation. We found that adult human beings had from 5 to nearly 500 milligrams of lead in their bodies, mostly in bone, indicating large differences in exposure, and that the average American had twice as much as the average African.

Industry made a "three-cities" study which showed conclusively that there had been no increase in air lead over a ten-year period in Cincinnati, Philadelphia, and Los Angeles, while a government study showed conclusively that air lead in 28 cities was directly proportional to the consumption of gasoline. Whom does one believe? Industry? When I said to one of their most avid proponents that traffic in those cities was already so dense at the beginning of the study that no more cars could squeeze in, he laughed and agreed that I was probably right—but he kept quoting those figures *ad nauseam*. Now they have "seven cities." They could have a hundred, but none with light traffic. The actual levels of lead in the air are shown in Table V-1, as well as the amount which could be retained in the body if the air were breathed 24 hours a day. In 10 years the amount from this source nearly equals the average body burden of American man, 120 mg. Industry says no problem! Lead does no harm.

In 1970 a meeting of 22 people was held in Puerto Rico to look into lead scientifically, what it does and what it doesn't do to human beings and animals. It was chaired by a professor of medicine, and all scientists interested were invited, except Dr. Hardy. When the list finally came out, it was loaded with oil men, including representatives from Ethyl. As nothing constructive could come out of such a group, I turned it down, having accepted on the basis of its being

Table V-1. Airborne Lead

Lead in the air of selected cities and some uncontaminated areas,
compared to the allowable limits in three countries.

City	Average	Min.	Max.	Heavy traffic	Amount retained in body at max. per year (milligrams)
	(micrograms [μg][1] per cubic meter)				
Cincinnati	1.4	1	2	14	7.3
Los Angeles	2.5	2	3	25	10.9[2]
Philadelphia	1.6	1	3	18	10.9[2]
Osaka, Japan	—	0.11	4.44	26	16.2[2]
Eleven towns, Japan	—	0.18	26	—	96.2[3]
Paris	—	3.25	9.8	—	36.0[3]
London	3.2	—	—	—	11.7[2]
Zurich	2.8	—	—	—	10.2[2]
Uncontaminated Areas					
Rural U.S.	0.5	—	—	—	1.8
Mountainous area, U.S.	0.2–0.01	—	—	—	0.7–0.04
Virgin Islands, Atlantic side	0.0	—	—	—	0.0
Allowable Limits					
Israel	5	—	—	—	18.3[2]
U.S.S.R.	0.7	—	—	—	2.5
Czechoslovakia	0.7	—	—	—	2.5

Note: People breathe 20 cubic meters of air a day. About half the lead in the air breathed stays in the lungs and is absorbed into the body. Most lead particles in air from car exhausts are very small and go deep into the lungs; larger particles do not, and are eventually swallowed.
1. One millionth of a gram.
2. Possibly toxic in 10 years.
3. Certainly toxic in 10 years.
SOURCE: Hernberg and Nordman 1972.

industry-less. How they muscle their ways in I don't know—money, I suppose. They have lots of it.

Industry is so virtuously concerned about the public health. We can't hold an impartial, scientific meeting without these watchdogs of power protecting their interests; their faith (and their pocketbooks) depend on "Lead is not harmful."

I have been interviewed on lead three times by television networks, each time for a scheduled 25 minutes. In each interview I pointed out the potential hazard of lead from gasoline to the public health, with data backed by experiments. The first, in May 1970 by NBC, was canceled—but shown in Australia, where you need a government permit to buy a can of lead-based paint. The second, by BBC in October 1971, was shown in full in England. The third, in December 1971 with the same camera and sound men as the first, was shown—but my remarks were emasculated and cut to less than a minute. I told the producer of each show "I must say some things the lead and oil industries won't like. I will bet they don't get past your sales force, who want Gulf to take Apollo to the Moon." BBC said they had no restraints and no axe to grind. NBC said both times that they had freedom to show what they wanted. NBC was naive, I hope.

So much for industrial virtue.

Some 600,000 young children living in slums are threatened with lead poisoning, and 50,000 to 100,000 are likely to require immediate treatment. Slum houses are old, and interior walls have been covered with lead paint—some of it half lead—which peels from the walls. The lead tastes sweetish and the children eat it. They have two strikes against them to start; airborne lead settles in the dust of city streets, which contains 0.2 to 0.5 percent lead, and in some cities, 2 to 6.8 percent. Playing children inhale or swallow this leaded dust. Additional lead from paint then poisons. The result at the least is mental retardation; at the most, death. Damage to the brain is permanent.

Nearly 2,500 cases of excessive exposure to lead were found in New York City in one year. Out-and-out poisoning is more frequent in the warm months, when children play in the streets, than in the cold. The surgeon general has begun a large campaign to find and treat such children, although little can be done to repair damage to the brain. Lead poisoning in children has been declared a medical emergency.

The Food and Drug Administration, a sometimes bumbling bureaucracy with its heart in the right place, sought to require warning labels on cans of paint containing more than a certain amount of lead. The Lead Industries Association fought back, and finally compromised reluctantly on a limit of 0.5 percent lead in paint. The American Academy of Pediatrics, which is concerned with the health of children, strongly recommended 600 ppm (parts per million), or .06 percent. Labeling only. Like cigarettes. And gasoline pumps. Some companies whose paint has no lead in it put the label on anyway, "to avoid lawsuits." The LIA is concerned about the health of our children?

The damage was already done, unwittingly, when the paint was applied, but children in the next century may benefit from this restriction. Not in this century. The cost of caring for those with lead poisoning has been put at 200 million dollars a year, for mental retardation and cerebral palsy.

Probably the most blatant example of industry's subtle and insidious pressures is the report of the National Academy of Sciences on health effects of airborne lead, which concluded it has "no known harmful effects." I quote from "News and Comment" in *Science*, November 19, 1971, which tells the story well:

A major report on the health effects of airborne lead, released by the National Academy of Sciences in September, has become the focus of a controversy over the academy's use of industry employees on its advisory panels to the Environmental Protection Agency (EPA). Critics in the environmental sciences community, including two prominent researchers who contributed to the report on lead, question the neutrality of the panel that wrote it and accuse the academy of giving scientists in the lead industry an excessively free hand in shaping the report, which was meant to serve as background for the EPA's regulatory policy on lead.

The academy, in turn, insists that industry is often the best source of essential expertise, and that when industry scientists serve on its advisory panels they are simply expected to rise above their allegiances to employers and to put aside their biases.

The report in question was written by an ad hoc Panel on Lead of the Committee on Biologic Effects of Atmospheric Pollutants (BEAP), a part of the National Research Council. The lead panel's report is the first of a series of similar surveys and evaluations of the literature on selected pollutants—which may eventually number as many as 20—being conducted by the NRC under contract to the EPA. Compiled over an 8-month period from July 1970 to February 1971, the report has been widely commended for its thoroughness in reviewing the literature on lead. The point of controversy is the panel's interpretation of the collected mass of information.

Early in its planning, the panel decided that, in order to place airborne lead in a proper perspective, it would have to expand the scope of its discussion and consider the effects of lead at far higher levels than those found in urban air. This was necessary, the panel said in its preface, "because lead attributable to emission and dispersion into general ambient air has no known harmful effects." From this premise, the panel worked its way through some 600 references to conclude that lead concentrations currently found in the nation's air pose no known hazard to the general population. Although the panel noted that some groups of workers and children in inner-city neighborhoods might potentially be at risk, it found that the amount of lead in the air of most major cities "has not changed greatly" in the past 15 years.

To judge from press releases issued in the wake of the NRC report, the lead industry was delighted with what it perceived as a clean bill of health from the National Academy of Sciences. The Ethyl Corporation, a major producer of lead additives for gasoline (the principal source of lead in the ambient air) took the report's conclusions as vindication of its contention that antiknock additives in no way "endanger the public health or welfare," and are therefore not subject to control on those grounds.

EPA officials, who had hoped that the report would furnish the scientific underpinnings for a national air quality standard to control lead (which would require evidence of a danger to health or welfare), showed considerably less exuberance. . . .

Eighteen scientists had a hand in writing the lead report's eight chapters and seven appendices. During their service on the panel, 4 of the 18 authors were employed either by the E. I. duPont de Nemours Company or by the Ethyl Corporation, which together produce most of the approximately 260,000 tons of lead additives burned each year in the United States. . . .

A contribution far more substantial than these, however, was made by Gordon J. Stopps, a duPont researcher whose views on the hazards of airborne lead have long been taken as representative of the industry's position. Stopps wrote the sixth chapter, which dealt with the role in air pollution of lead additives themselves, as opposed to their combustion products. He also wrote the report's discussion of "epidemiology of lead in adults," a topic of particular interest to the EPA. Stopps has since left duPont and moved to the Canadian Department of Health in Toronto.

The first complaint about the author's affiliations was lodged with the academy by Harriet Hardy in the summer of 1970, shortly after the panel first convened. Long associated with M.I.T. and Massachusetts General Hospital, she was one of two preeminent figures in the field of metal poisoning who were asked by the academy to serve as anonymous, outside reviewers of the lead report as it progressed through several drafts. The other outside reviewer was Robert A. Kehoe, an emeritus professor of the University of Cincinnati's Kettering Laboratory. A medical consultant to the Ethyl Corporation since the late 1920's, Kehoe had the distinction of being cited in the lead panel's list of references a dozen times, more than any other researcher.

In her letter to the academy, Hardy protested that the list of authors was "top heavy" with industry scientists in general and with Stopps in particular. Tsaihwa J. Chow, a research chemist at Scripps Institution of Oceanography at La Jolla, and a contributor to the report's first chapter, filed a similar complaint this past August, soon after he learned of

Stopps' contribution. Chow said that even if Stopps were not biased, his association with the panel as an industry scientist would damage the report's credibility.

Both Hardy and Chow were told that it was academy policy to draw experts from wherever they might reside, without regard to affiliation. Elaborating on this policy, [Louise] Marshall [NRC staff officer] told *Science* that the issue of conflict of interest was discussed in the first BEAP committee meeting in the spring of 1970. She said academy officers "made it clear that committee and panel members were being asked to serve as scientists and not as representatives of their organizations." Moreover, she said, it was the academy's experience that industry scientists "tend to lean over backward to be fair."

Hardy and other critics of the lead report remain unconvinced, however, that declarations of neutrality mean very much. "How could he be neutral?" she said of Stopps, in a telephone interview. "He has written and written for years that there's nothing harmful about tetraethyl lead. . . . It's just not possible for him to act purely as a scientist." She said that the academy seemed naive in its use of industry scientists and noted that "it's a queer thing that they haven't learned what other government agencies have about the objectivity of industry." . . .

An academy public affairs officer said that seemingly the most effective means of compensating for the biases of advisory panels is simply to strive for a balance of opposing philosophies. As he expressed it, "Whether this is naive or not, the feeling is that, if a panel is balanced, then a person with any conflict of interest will not carry undue weight."

In Hardy and Kehoe, the academy struck something of a balance between left- and right-wing views of lead pollution. But no such balance was apparent within the panel itself; in the opinion of members who were contacted, there was no identifiable "environmentalist" among them who might have served as a counterpoise to industry's weight. . . .

Apart from questions of its membership, the lead panel has given rise to accusations that it had a pro-industry bias—or at least the appearance of one—by virtue of its curious selection and treatment of data from an important government-industry study of lead in the air of seven U.S. cities. The panel only indirectly acknowledged the the existence of one set of data from the study, which runs counter to the panel's conclusion that airborne lead concentrations have not, for the most part, changed greatly in 15 years. But the panel made explicit reference to other data from the same study indicating that—as the lead industry contends—no meaningful relation exists between lead in blood and lead in the surrounding air. . . .

The other set of data released in November showed that, during the 1960's, airborne lead increased at 17 of 19 measuring stations in the three cities described. Increases at Los Angeles sites ranged from 33 to 64 percent above values measured at the same sites in 1961–62;

from 2 to 36 percent in Philadelphia; and from 13 to 33 percent in Cincinnati. In its most direct reference to this information, the academy report states that:

"Preliminary data on samples taken in 1968–69 from the same sites as in 1961–62 indicate that air lead concentrations at some individual sites are higher than in 1961–62." . . .

When the old National Air Pollution Control Administration (now part of the EPA) signed its contract with the academy last year to procure advice on pollutants, it did so in order to counter industry criticism that the background papers it was using to justify ambient air quality standards were faulty. In turning to the academy, it may be that conservatism was the price the old NAPCA and the new EPA paid for credibility. Nevertheless, EPA officials say they still feel free to accept or reject the academy's interpretation of the literature it compiles. As Irwin Auerbach, of the Office of Air Programs puts it, "We can use the bulk of the lead report or we can reject it if we feel it's slanted. But bias is not the sort of thing you expect to come from the National Academy of Sciences."—Robert Gillette.

How anyone can believe that a man can write an impartial summary opinion when his job depends on his taking a biased viewpoint for industry is beyond me, especially since his opinionated bias has been well known for years. The report is good but the conclusions are incredible! Not only have duPont and the oil industry lost what credibilities they had, but the illustrious National Research Council, long known as a hotbed of inactivity and conservatism, has lost its prestige.

Some of us remember that it was another committee of the National Research Council-National Academy of Sciences which took 30 to 35 years to recognize that Vitamin E and Vitamin B_6 had a place in human nutrition, in spite of requirements for other mammals having been established. Did they believe subconsciously that man is so special that he differs from other mammals in not needing the essential nutrients all mammals need? More likely they could not agree on a number, the amount probably needed, and so ducked the issue, thus indirectly depriving a hundred million people or more— growing children, adults, and aging persons—of vitamins in amounts adequate for optimal health. Vitamin E is lost in storage and freezing, and B_6 by heating.

A committee, by its very nature, seeks the highest level of inactivity and is dominated by its most vociferous members. In the lead report, duPont dominated the committee's conclusions, which were slanted toward doing nothing, not changing the status quo, and allowing duPont and Ethyl to continue polluting the environment with lead. Just to be safe, this committee had a representative of the paint industry. Five members with axes to grind are enough to swing 13 others, many of them with few opinions and scanty knowledge of the whole problem in human terms.

What are the facts?

Lead is being extruded into the environment from the tail pipes of motor vehicles at a yearly rate of about two kilograms (4.4 lbs) per car. It lingers long. It is breathed by drivers, passengers, and people living in cities and near heavy traffic, who absorb it through their lungs. It accumulates in the body with age. It is a protoplasmic poison, not a very strong one in small amounts, but there is so much around that it has an effect on everyone exposed to present environmental concentrations—a measurable effect.

Mice and rats fed lead for life in concentrations reproducing the levels in American tissues—man's unwitting experiment on himself—show early mortality, shortened life-spans, shortened longevities (dying before ripe old age), susceptibility to infections, visible aging, and loss of weight. When breeding rats were exposed to the same concentrations, the strain was dead in two generations (like the ancient Romans). In addition, there were many heart attacks and much hardening of the arteries. Applied to man, these findings would make lead the number-one public health problem. Some of these subtle effects were partly prevented by enough chromium in the diet—but American bodies are generally low in chromium.

Patterson was right; 800-year-old Peruvian Indian bones had no detectable lead, and there were only very small amounts in third-century Polish bones (although later there was much, probably from pewter and lead utensils). There is little or none in most children's bones even today, except slum children, and none in bones of the newborn.

In Sweden, estimates are that lead in the air will increase 100 to 200 percent during the next ten years. On this basis, and the obvious threat, Sweden is phasing lead out of gasoline. The oil and lead interests apparently are not as powerful in Sweden as they are in the U. S.

Not all the tetraethyl lead in gasoline is burned and exhausted into the environment. Estimates vary from 50 to 75 percent in exhaust gases, with 50 to 25 percent remaining in the engine oil. Discarded engine oil used to be re-refined for lubrication, more than 500 million gallons being drained from crankcases per year. This oil contains more than 55 million pounds of metal oxides, of which 32.5 million pounds is lead. A letter to the author from V. T. Worthington, Executive Director of the Association of Petroleum Re-refiners, dated Jan. 6, 1971, states:

> The re-refining industry recycles about 150 million gallons of drainings each year but because of discriminatory taxes and labeling levied against this industry by the government, it is not economically feasible to recycle all the oil drainings available. Encouraged by high fuel oil prices and recommendations of major oil companies, more and more of the drainings are being burned. Because present fuel oil specifications do not take into account that metals are present, there is no law to prevent this burning except the pollution control laws. Our concern is that by the time the government gets around to monitor with equipment that will pick up the contaminants, irreparable damage will have been caused and the recycling industry may not be around to perform their valuable service.

In that event, all the lead added to gasoline will end up in the air, instead of only 85 to 92.5 percent of it. That is a high price to pay for jack-rabbit starts. The oil lobby is powerful around Washington.

What is suspected? A number of medical scientists suspect, but have proven in only a few cases, that some vague and ill-defined human ills common in the urban population are manifestations of mild lead poisoning. Lead goes to the nervous system. That tired, run-down feeling, nervousness, depression, apathy, lack of ambition, frequent colds and other infections, and mild psychoneuroses may be the result of lead in the air.

Most of the body's lead is deposited in bone. Nearly 50 years ago it was shown that bone lead would pour out when the body was under physical stress. We don't know what effect this additional poison might have on a person recovering from an accident, an operation, or pneumonia or other infection, but it could hardly aid recovery. It might tip the balance between life and death.

When the soft tissues of people dying of a variety of diseases apparently unrelated to lead intoxication were compared with tissues of persons dying of accidents, it was found that there was several times as much lead in the former group as in the latter. Thus the evidence is indirect but suggests that extra lead may have influenced the fatal outcome of the diseases.

Every urban dweller is exposed to lead and carries it in his blood. In persons slightly poisoned by lead, ALA (delta-aminolevulinic acid) is found in the urine as a result of inhibition of an enzyme system concerned with the synthesis of hemoglobin. This enzyme, ALA dehydratase, is extremely sensitive to lead. Even at blood lead levels so low that the poisoning is not enough to detect ALA in urine, inhibition of this enzyme occurs. And inhibition is proportional to lead in blood. It is hardly comforting to know that all of us who drive are slightly poisoned, some more than others, although we don't know it.

We have long recognized the bad effects on the human body of too much ionizing radiation and have contrived stringent regulations to control it. In fact, we have become almost paranoid about it— witness the strident cries of those with little knowledge trying to hamstring the construction of nuclear power plants. At the present rate of uncontrolled emissions, lead represents a far greater threat to human health than do power plants and reactors.

Lead is a most useful industrial metal and has been for thousands of years. There is little on the surface of the earth, only 10 to 14 ppm, which means that in a ton of rock or dirt one could find only 10 to 14 grams of lead. Some unpolluted soils have only 2 grams per ton. In contrast, there are 56,300 ppm, or 56.3 kilograms (124 lbs) of iron in a ton of igneous rock, and 82.3 kg of aluminum. Since living things have been exposed since the beginning to such small amounts of lead, they have not needed to learn to handle it. Therefore it is toxic in large amounts.

Lead is deposited largely underground. Man has mined it and has spread it over the surface of the earth, during the past 200 years from emissions of smelters and during the past 50 from emissions of motor vehicle exhausts in order to improve the performance of his cars. But man has other exposures to lead. Table V-2 lists them. Fortunately, he absorbs only 5 to 10 percent of the lead he eats, average 7 percent, unlike the 30 to 50 percent he absorbs from the air he breathes. Soft acid water dissolves lead from pipes and has caused lead poisoning. There is not much lead piping left in houses in the U. S., except between the water mains and the houses in old cities, and on old farms. All but a very few large municipal water supplies have tolerable levels of lead; so do all but a few major rivers and lakes. Food grown near highways is often badly contaminated and could be toxic, especially lettuce and potatoes. Hay polluted with airborne lead has killed a horse in Wales and seven Apaloosa horses in California.

Table V-2. Some Sources of Lead Pollution

Uses of lead with potential or actual contact with human beings, thousands of short tons, U.S., 1967.

Use	Thousands of tons	Pollution of	Remarks
Antiknock gasoline additives	247.2	Air, soil, water	Nearly all in air
Storage batteries	237.0	Air and water	Fraction burned
Paints and pigments	103.0	Water and soil	Released by weathering and burning
Ammunition	78.8	Water and soil	In wasted shot. Also enters body via wounds.
Solder	68.8	Water	In pipes
Caulking compounds	48.8	Water	In pipes
Type metal	28.6	Air	When cast
Brass and bronze	20.5	None	Probably inert
Pipes and traps	20.0	Water	Waste water
Collapsible tubes	11.0	Little or no pollution	None in toothpastes
Lead foil	6.2	Food	Direct contact
Annealing	4.2	Air	When burned
Galvanizing	1.9	Water	When corroded
Plating	0.5	Water	When corroded
Total metal products	876.5		
Total domestic production	943.0		
Imports	488.0		
Total	1,431.0		

SOURCES: Hernberg and Nordman 1972, and *Mineral Facts and Problems*.

Some lead gets into food from processing, but today most canned food does not come into contact with lead solder. Lead salts are added to some Italian wines to stabilize them, but not to American or French wines, and some lead enters the wine from the lead foil about the neck of the bottle. Cheese and chocolate wrapped in lead foil absorb lead. However, toothpaste does not seem to absorb lead from collapsible tubes. Pewter, britannia metal, and solder contribute to our intakes when they are in contact with foods. At the Trace Element Laboratory we found much lead in lobster, yeast, puffed rice, organically grown kale and lettuce, dry milk in packages, tea, cider, vinegar, and dog food. Lead arsenate has been used as an insecticide since 1895, and there is enough lead from this source to contaminate apples, grapes, and tobacco—there is some in cigarette smoke (20 micrograms per pack). As said before, there is enough lead in grass growing along a secondary highway opposite Brattleboro Memorial Hospital to abort a cow, and the amount has quintupled in 10 years; there is no detectable lead in a pasture five miles away. (Let the oil companies explain that!)

Rats inhale so much lead from discarded batteries burning in dumps that they get kidney tumors. People who work with lead—painters, plasterers, caulkers, plumbers, type setters, solderers—are poisoned fairly often although industrial standards are quite rigid. Lead bullets have caused poisoning after many years in the body. I talked to a bedridden woman who had become partially paralyzed as a result of permanent lead poisoning. For six years she had taken a dietary mineral supplement bought at a "health food" store.

There is an old medical case of a workingman in England who came down with lead poisoning. Doctors searched for the cause but could find none. Finally, a bright young doctor followed him around all day; it was his habit to stop at his neighborhood pub for a pint of bitters every morning before he went to work. He was usually the first one in the pub and was served the first mug of draught beer. The doctor examined the beer pump and found a short length of lead pipe between the pump and the keg. Standing overnight in the pipe, the beer dissolved enough lead to cause poisoning. No one else was affected.

Every year a number of rural people in the Deep South suffer from lead poisoning and several die. The source: illicit moonshine whiskey distilled in old automobile radiators. Children are poisoned by sucking lead-painted toys or lead soldiers, or lead-pigmented crayons, or by eating old putty from windows. The glaze on some Italian pottery is dissolved by acid fluids and gives up much lead; children have been poisoned by drinking orange juice left standing in fancy earthernware cups.

The lead industry is threatened today in many of its aspects. Red lead is an excellent rust preventive for steel and iron, and will probably continue to be used in the U.S. in annual amounts of 80,000 tons or so, but white lead is being replaced by zinc and titanium; the 26,200 tons of lead pigments are not entirely necessary. Substitutes for the nearly 50,000 tons of lead in putty and caulking compounds are available, but lead solder is cheap and useful. Incursions into type metal are already deep through photographic techniques and computerized composition. Waste pipes and traps of lead are giving way to plastic, as are many collapsible tubes. Lead foil is being replaced by aluminum, which is nontoxic, and lead galvanizing by zinc. The use of long-lasting and efficient cadmium-nickel storage batteries in automobiles has long been opposed by the powerful LIA in favor of the lead battery with a life of only two years; obviously a loss of 16.6 percent of annual sales of lead would be bad for business. Another loss of 17.3 percent in sales of lead added to gasoline would be almost disastrous. The lead industries are fighting for their corporate lives and profits, and it is no wonder they resort to infighting, subterfuge, false propaganda, subversion, and other tactics described by Mr. Winn.

There has been an "octane race" in the last 15 years, with larger and more powerful motors burning higher and higher octane gasoline. Fortunately, it is over. "Regular" gasoline without lead has a higher octane rating today than leaded gasoline had ten years ago. The oil companies can produce high-test gasoline without lead—one of them does.

Lead gums up the pollution-control devices which will be mandatory on all cars. For this reason the government is advocating its removal from gasoline. Lead in the air gums up certain functions of the human body. For this reason it should be removed from gasoline. If anyone tells you that lead in the air "poses no hazard to human health" your answer should be "That's ridiculous!" Lead is everywhere, and it is everywhere in increasing quantities. The least we can do is to keep it out of our lungs and the brains of our children.

An estimated one to five million children are "hyperactive," that is, they are too restless to be quiet at school or at home. They are so constantly moving that they cannot concentrate on anything. A recent report on a considerable number of such children showed that they had moderately elevated levels of lead in their blood and excreted larger than normal amounts when given a deleading drug. So another former hidden manifester of lead poisoning appears.

A child playing in the street drops its lollipop and eats it, or gets its hands dirty and puts them in its mouth. A few such performances can give it lead poisoning from which it never recovers. We dropped some lollipops and picked up as much as a milligram of lead on Main Street, Brattleboro, Vermont.

Because of the threat to children and the finding of large amounts of lead in the dust from the streets of 77 cities, 49 members of Congress (none from oil-producing states) signed a letter written by Senator Philip A. Hart to William P. Ruckelshaus, then director of the Environmental Protection Agency, asking that all lead be removed from gasoline by 1977. Present EPA plans call for a reduction of emissions to 50 percent. The letter states that that is not enough. And it isn't. After all leaded gasoline is banned it will take a hundred years for our cities and highways to recover from our gigantic splurge of pollution.

If rural America takes comfort in the idea that small towns are free of this hazard, rural America is in for a rude shock. In our small Vermont town, there is as much lead in the air and the dust of Main Street as in New York or Chicago air and dust. Wherever the car goes, two pounds of lead per man, woman, and child comes out of its tail pipe every year, or about four pounds per car. Where does it go? Some of it goes into children. See Tables V-3 and V-4.

Monitoring agencies have one bad habit which leads to a false set of values for lead in air. They measure it at heights where apartment dwellers would breathe it, not at street level, where children playing or pedestrians would breathe it. This habit has led those who think about lead to believe that there really isn't very much around, as the "seven-cities" survey would have us believe. Everyone has fallen into this trap, including the EPA administrator himself. To show its falsity, here are some figures of analyses made on air at different levels above Main Street, Brattleboro, a town of 14,000 persons in micrograms per cubic meter ($\mu g/m^3$): 1 ft., 10.4; 2 ft., 14.3; 3 ft., 9.0; 5 ft., 8.3; 8 ft. (too high except for a man on stilts), 2.8; 30 ft., 5.1. I believe that none of the increased lead levels in the air of cities represent what we actually breathe.

There are many other sources of human exposure to lead (Table V-5) and we must not forget them. But only 5 to 10 percent of the amount we eat or drink is absorbed into our bodies, versus 50 percent of what we breathe. Most of the sources listed are eaten or drunk, and it takes quite a bit of lead to poison us that way. Over 90 percent of our lead comes from gasoline.

Table V-3. Lead in Street Dirt of 77 American Cities

State	No. of cities	Lead in dirt of streets	
		Residential areas	Commercial areas
		(parts per million)	
Arkansas	1	1,163	2,775
Colorado	3	1,741–2,974	2,843–4,065
Illinois	9	656–3,067	1,774–3,549
Indiana	8	290–9,972	942–6,597
Iowa	5	849–2,525	1,127–3,790
Kansas	3	865–1,558	1,498–2,080
Kentucky	4	1,945–2,702	1,350–5,089
Michigan	9	1,041–3,042	2,524–4,722
Minnesota	3	1,265–2,563	1,650–4,681
Missouri	3	1,566–4,681	2,086–4,639
Nebraska	2	1,891–2,603	2,256–4,464
North Dakota	1	958	2,522
Ohio	12	206–2,639	352–2,933
Oklahoma	2	1,683–1,950	2,272–4,935
Tennessee	4	1,010–4,416	2,677–20,667
West Virginia	5	1,084–2,045	375–6,979
Wisconsin	3	912–1,993	1,895–3,331

SOURCE: Analyses by Environmental Protection Agency.

This story could have an unhappy ending, for things moved fast. On November 29, 1972 there appeared EPA's "Position on the Health Effects of Airborne Lead." It is a good document, which cries out repeatedly (but does not actually say) that the sources of lead must be reduced to a minimum. Table V-4 has some of the data, showing that one out of four children is threatened with lead poisoning.

The *Federal Register* of January 10, 1973 carried a proposed "Regulations of Fuels and Fuel Additives." One part dealt with lead as a poison for emission control devices. The solution: (*a*) All cars with emission devices shall be labeled "Unleaded gasoline only." (*b*) All large gas stations shall have one or more pumps with small nozzle spouts for unleaded gasoline of 91 octane rating. Spouts for leaded gasoline shall be large. (*c*) Cars taking unleaded gasoline shall have

Table V-4. Incidence of Overexposure to Lead

Proportion of children aged 0–5 years and adults in various cities having blood
levels of lead considered indicative of overexposure (40 micrograms per
100 milliliters)

City	Children 0–5 yrs. (%)	Adult males (%)	Adult females (%)
Baltimore	25.3–31.5	—	—
Chicago	20.0	—	—
New York	20.2–45.5	—	—
Philadelphia	34.0	2.3–4.5	0.7
Washington	22.0–39.2	—	—
New Haven	23.7–29.8	—	—
Newark	38.9	—	—
Norfolk, Va.	22.7	—	—
12 Illinois cities	11.4–31.3	—	—
Cincinnati, various occupations	—	2.9–67.0	—
Los Angeles, various occupations	—	0.6–5.2	3.3–4.4
Oakland	—	5.5	1.9
Camden, N.J.	—	—	1.8
Residence near a roadway	—	—	1.8
125 feet from a roadway	—	—	0
400 feet from a roadway	—	—	1.6
Average	24.6	2.76	2.2

Note: The number of children actually threatened by airborne and dustborne lead
from gasoline additives is considerably greater than shown here. Children have
fewer red blood cells than adults, and these figures were not corrected for this
difference. About 90% of the lead in blood is in the red blood cells. Correcting
for this factor would increase the percentages of children at risk by 15% to 28%.
Source: Environmental Protection Agency, "Position on the Health Effects of
Airborne Lead."

small filler inlets to fit the small spouts, those taking leaded gasoline
shall have large inlets. All new cars are supposed to have emission
control devices, but the kind is not yet settled. Thus, the owner of a
new car will be forced to use lead-free gasoline.

The second part of the proposal admits the danger from airborne
lead and seeks to lower average air levels to 2 micrograms per cubic
meter or less. (Table V-3 shows the levels in some cities; note that
Philadelphia would be clean under this ruling, although one out of

Table V-5. Sources of Lead Other than Leaded Gasoline

Sources potentially or actually contributing to increased body burdens of lead and occasionally causing poisoning.

Source	Means of contamination	Means of entering body	Result
Paint, inside	Direct	Eaten by small children	Poisoning, mental retardation
Plaster and putty	Direct	Eaten by small children	
Lead toys, crayons	Direct	Sucked by children	
Lead glazes on pottery	Acid foods and fluids	Ingestion	
Old auto radiators	Illegal whiskey distilling	Drinking	Several deaths annually
Bullets or shot	Tissues	Wounds	Slow to develop poisoning
Lead pipes	Soft acid waters	Drinking	Long use necessary for poisoning
Lead pipes	Beer standing overnight in keg	Drinking	Only first glass produced poisoning
Shotgun pellets	Direct	Eaten by ducks	Ducks poisoned
Lead amphorae	Wines, syrups	Drinking	Probably poisoned ancient Romans
Wines	Lead cork covers	Drinking	
Whiskey	Distilling	Drinking	
Paint, outside	Water, soil	Food, water	Absorbed by vegetables
Plastics	Air, by burning	Lungs	
Foundries, battery makers	Air, water, soil	Food, water, air	
Used motor oil	Air, by burning	Lungs	
Insecticides	Soil	Food	
Cigarette smoke	1 μg per cigarette	Lungs	
Weighted clothing	Skin	No absorption	

three children is threatened at a lower level.) Its avowed aim is "to reduce preventable lead exposures from automobile emitted airborne lead to the fullest extent possible." To achieve this pious hope, in the face of powerful interests to the contrary, the EPA administrator proposed to cut the lead in gasoline in half, progressively to 1978!

This move is not "to the fullest extent possible." If the consumption of leaded gasoline were to double in the next ten years, we would be where we are now, but worse off. The fullest extent possible is to eliminate lead completely. But the vested interests must be placated, while the costs of caring for permanently damaged children continue. This compromise might be acceptable if it were impossible to produce a lead-free gasoline with an octane rating high enough for our large cars. But it is not. Today's high-test lead-free gasoline will power a Cadillac. In properly designed engines, lead additives have little or nothing to do with gasoline mileage. Today, lead-free gasoline has as high an octane rating as did leaded gasoline in 1960. It is the high-powered cars that demand additives to most, but not all, gasolines.

I brought a Citroën from France. It was designed to run on Spanish gasoline of 45 octanes up to American gas of 90 octanes. Once it was inadvertently filled with high-test gasoline. It choked and sputtered and would not run. I had to drain the tank. At long last, the American automobile industry is turning to more economical engines, using lower octane gasoline.

There is one bright spot on the horizon. New York City banned leaded gasoline as of July 1973. Ten major oil companies have protested vigorously, but Jerome Kretchmer, the city's former environmental protection administrator, stuck to his guns on the basis of the *threat of lead to the public health*. It is high time the federal government followed suit, and stopped compromising. If lead is a health hazard, it should be banned. The problem is as simple as that.

A shortage of petroleum in the 1970s was predicted over 40 years ago, and was certain to become severe because of recent levels of consumption and the annual increase of five to six percent in the use of oil and gasoline. It would not be surprising to learn eventually that the "shortage" of gasoline in the summer of 1973 was engineered deliberately by the oil interests to raise prices and to force the independent gas stations out of business. The 1973 Arab oil embargo merely precipitated the present "crisis." It would be foolhardy to allow that "crisis" to scare the public into forgetting the health hazards of lead additives under the threat of shortages and rationing.

VI

MERCURY AS A CAUSE

OF NATIONAL PARANOIA

Mercury as element no. 80 in the periodic table is one of the 81 stable elements in the universe. As such, it is everywhere—in rocks, soils, plants, animals, water, air. It is a liquid at ordinary temperatures, and those of us who live in cold climates know that when the mercury freezes, it is 42 degrees below zero. When the weather is warmer, mercury vaporizes a little, so that there is always some in the air, coming from the ground. Rain contains mercury. Too much mercury can produce mental changes in man.

Mercury was named for the Roman god, the messenger of the gods, the god of commerce, manual skill, eloquence, cleverness, travel, and thievery. Quicksilver is elusive, heavy, hard to capture, volatile, changeable, fickle; when you spill it, it breaks up into a thousand tiny globules and virtually disappears. Mercury combines immediately with gold and silver—watch your dentist prepare a filling for a cavity. It likes sulfur, and most of the mercury around is combined with sulfur, in the ore cinnabar.

Mercury has many uses. Formerly one of its largest was as an extractor of gold. Crushed gold ore was passed over a plate covered with a film of mercury. The amalgam was scraped off and heated, and the mercury boiled off to leave the gold. A flask of mercury, the unit of measurement, weighs 76 lbs; a burro can carry two flasks, one on each side, into the mountains for gold refining at the site.

At the present time, mercury does many things useful to man (Table VI-1). It keeps his paint and his seeds from mildewing, fills his cavities, catalyzes salt to make soda and chlorine gas, catalyzes gases to make plastic, disinfects, helps make his paper and pulp, and has hundreds of applications in electrical and laboratory apparatus. When he wore hats, it helped make the felt.

Being expensive, $10 a pound, mercury is usually recovered from most processes, but from others it cannot be saved, and it is dumped into water as used catalyst. Whether dumped or recovered is a matter of dollars and cents (fractions of a cent)—if it costs more to recover than to buy it is discarded. But not much is actually dumped into rivers and lakes, a few pounds here and there.

Nature weathers and transports to the oceans every year about 3,500 tons of mercury from natural sources—2,500 tons dissolved in the rivers of the earth and 1,000 tons carried in sediment. About one ton remains dissolved in seawater, the rest being precipitated to the bottom. There are 1.42 quintillion (1.42×10^{18}) tons of seawater, and seawater is an excellent scavenger of metals.

In the United States, about 655.5 tons are *potentially* capable of being dumped into rivers and lakes every year from chlor-alkali plants, other catalytical processes, and papermaking. What fraction is actually dumped is not known; probably all from papermaking and some from catalysts and chlor-alkali plants. That mercury can be recovered. In addition, 141.7 tons were formerly spread over the ground in seeds and fungicides and largely stayed in the soil. When mercury gets into soil, it seeks out sulfur and becomes insoluble. When it accumulates in lakes and rivers, most of it is bound firmly to sulfur, and in that form it is largely unavailable to living things.

There is a prevailing belief that we have polluted our waters with mercury because of the nefarious carelessness of the industries that use it. However, *not one* of the 720 major rivers and lakes in the United States examined by the U.S. Geological Survey contained mercury in amounts exceeding the official limits for potable water, 5 parts per billion. (See Table VII-1.)

Furthermore, it is clear that human activities cannot yet have had a significant influence on the amounts of mercury in the marine environment, whatever the effects on local environment. In the English Channel and off the coast of England, seawaters have had 0.01 to

Table VI-1. Sources of Exposure to Mercury

End uses of mercury having potential or actual contact with human beings, U.S., 1968.

Use	Metric tons	Environmental exposures
Electrolytic preparation of chlorine and alkali	543.78	Can be discharged into lakes and rivers.
Paint, mildew proofing and antifouling	239.4	Evaporates into interior air in small amounts. Some was methyl mercury.
Agriculture, fungicides and bacteriocides	141.74	Soil and water. All remains in soil. Much was methyl mercury.
Catalysts	94.62	Can be dumped into lakes and rivers. Methyl mercury often formed in process.
Pharmaceuticals	74.1	On or into human beings.
Dental preparations, fillings	51.68	In teeth.
General laboratory use	42.94	Laboratory workers.
Paper and pulp manufacture	17.1	Dumped into rivers and lakes.
Electrical apparatus	506.18	Sealed. No exposure.
Industry and control instruments	147.06	Sealed. No exposure.
Amalgamation	83.22	Jewelry.
Other	477.28	—
Redistilled	278.54	Used again.
Total consumption, including redistilled	2,697.64*	
Total domestic production for export	803.7	
Total released from stockpiles	694.64	
Imports	908.2	
Nature's weathering to the seas	3,500.0	
Man's burning of fossil fuels, to the air	3,000.0	

* In 1969 total consumption was 2,729.09 metric tons.
SOURCES: U.S., Dept. of Interior, Geological Survey Professional Paper 713; and Wallace et al. 1971.

0.021 parts per billion, and English rivers 0.009 to 0.01 ppb. Atlantic seawater at various depths up to three miles had 0.003 to 0.02 ppb, with the lower levels at the greatest depths. Thus, there is no evidence that the oceans or rivers are polluted with mercury.

As one might expect, bottom mud has contained larger amounts, 0.02 to 1.0 ppm, a result of accumulation from the land. These values, however, are within the range of the abundances in most rocks on land, and are not alarming, though they are higher than in some soils. Table VI-2 shows the average and extreme concentrations of mercury in air, soil, rocks, and water. Variations in the levels of mercury from place to place were very wide, except in water.

Coal contains varying amounts of mercury, and every year 3,000 tons enter the atmosphere from this source. Some crude oils have considerable mercury, and when burned raw could put many hundreds of

Table VI-2. Mercury in the Natural Environment

	Minimum found	Range of means[1]	Maximum found
Ocean water, ppb	0.006	0.014–0.021	0.27
Ocean sediments, ppb	0.0	80–1,200	2,000
River water, ppb	<0.1	0.2–2.8	3.5
River sediments, ppb	trace	1,000–2,700	26,000
Lake sediments, ppb	360	585	4,000
Rocks, all types, ppm	2	5–5,000 (80–400)	30,000
Soils, ppb	10	23–1,300 (30)	10,000
Coal, ppb	1.2	12–46,000 (3,300)[2]	300,000
Petroleum, ppb	1,000	—	500,000
Air, $\mu g/m^3$	0.6	4.5–190	300
Rain water, ppb	0.05	0.2	0.48
Plants, ppb	<500	150–300	3,500

Note: Data from areas of mercury deposits were omitted, except one for California petroleum.
1. Numbers in parentheses are from Bowen 1966.
2. Average of 26 U.S. coals.
SOURCE: U.S., Dept. of Interior, Geological Survey Professional Paper 713.

tons into the air; others have little. When petroleum is refined, the mercury stays in the residual oil, or if heated, escapes; there is no mercury in gasoline and probably none in diesel oil.

The mercury industry, unlike the lead and petroleum industries, has no powerful lobby and no vociferous spokesmen. It is a small industry dealing in a few thousand tons, instead of in a million or more. It must sit quietly by while the false cry goes out that it is polluting the earth and the seas. And poisoning man.

Man, himself, our Reference Man, or average American, has 13 mg mercury in his body, of which his brain, kidneys, liver, and hair contain the most concentrated amounts. There are 4.2 mg in muscle, 4.5 mg in fat, 1.4 mg in the brain, 0.9 mg in the kidneys, 0.5 mg in the lungs, and 0.5 mg in the liver. The concentrations in his hair (6 ppm), kidneys (2.8 ppm), brain (1.0 ppm), and lungs (0.58 ppm) exceed bureaucratic limits for food (0.5 ppm). To kill him takes 2.2 to 22 grams of mercury as mercuric chloride by mouth.

When man ingests or breathes mercury, it leaves the body fairly quickly, half being gone in five days, unlike lead and cadmium, which stay. It takes quite a bit of mercury every day, about 100 mg, to make him sick. But if he keeps taking sizeable amounts, some collects in his kidneys, fat, and brain. Ordinary amounts, perhaps 0.02 mg per day, he gets from food, water, and air. Fish have quite a bit of mercury in their flesh, especially fatty fish.

Probably all living things have contained mercury since the beginning of life, but in very small amounts. Fish are no exception. More mercury was found in bones from prehistoric fish than in bones of present-day fish. Mercury is not a fish pollutant, except under special circumstances. Preserved fish caught 40 years ago have shown a lot of mercury in their flesh. Mercury accumulates in fish with age, the larger fish having more. Fish-eating fish have more mercury than clam- or plant-eating fish. In fact, all fish-eaters, be they fish, flesh, or fowl, have more mercury than flesh-eaters or grass-eaters or seed-eaters— unless the seed has been soaked in mercury to prevent it from molding.

We are especially interested in the question of whether or not fish are safe to eat, because recent rulings by the Food and Drug Administration have ruined the swordfish industry, hampered the tuna industry, curtailed sports fishing in several states, and enhanced a wave of pollution paranoia across the nation. The decision to impound fish with more than 0.5 ppm mercury was the result of a discovery by the press that fish contained mercury—as they always have—and an assumption that all mercury is poisonous. It followed an action by the Swedish government to set the limit of 1.0 ppm mercury for edible fish. Both decisions were made without critical examination of the facts, and both assumed that man has polluted the oceans.

This assumption is ludicrous, as anyone knows who has studied the facts. If all the mercury produced in the world every year, 10,000 tons, were dumped down the drain, and all of it remained dissolved in the ocean, 10,000 tons in 1.42 quintillion tons of water, there would be 7 parts per quadrillion added annually. But as mercury is scavenged by sea organisms and drops to the bottom, there would be only three tons in solution, making an increment of mercury in seawater of 2 parts per quintillion. Neither of these increments could be measured in a thousand years. Therefore the conclusion is inescapable that whatever mercury is found in ocean fish is natural, or background, mercury, which has been there since the year one. And furthermore, any dictum putting a limit on what natural background levels are bad for us is a bureaucratic encroachment on human rights and on Nature, and should be illegal.

So much for ocean fish, swordfish, tuna, cod, halibut, mackerel, and the like. But a letter from the late Nick Begich, congressman from Alaska, states that Alaskan ocean fishermen are complaining heavily; they have to measure each fish and throw back those over the size believed to contain more than the allowable level of mercury—although the mercury is not measured—and that takes so much time that their profits are endangered.

Fish have an unusual ability to accumulate surprisingly large amounts of mercury from the very low concentrations in seawater—or in fresh water—as we are beginning to realize. Mako sharks from Long Island waters have 7 ppm; these sharks are a delicacy on the table. If one puts soluble mercury into an aquarium, the fish will build up as much as 20 ppm mercury in a few weeks. And they lose it slowly, unlike mammals, who lose it rapidly.

We do not know if mercury in the bottom muds of estuaries affect levels in fish, but from the scattered reports it does not seem to. Only when the mercury is dissolved in water does it seem to enter the food chain.

At the time of the Great Mercury Scare, 51 industrial firms were discharging mercury into rivers and lakes of the United States in amounts of 0.05 to 60 pounds per day. Nine months to a year later, the amounts were markedly reduced: 19 firms each discharged less than 0.5 lb, and discharges had been reduced substantially by 29 others, of which only six discharge more than 1.0 lb per day. In the cases of the remainder, discharges were very low. Despite the fact that the dumping was rapidly ceasing and that little mercury was found in water, rivers and lakes in 18 states were restricted for fishing. Officials were fearful that mercury in sediments would get into the food chain.

That mercury dissolved in water does get into the food chain is demonstrated in Table VI-3. Considerably higher levels were found downstream from paper mills and chlor-alkali plants than were found upstream. A paper mill discharging water-soluble phenylmercuric acetate, which has low toxicity, was said to have caused large amounts of methyl mercury in fish downstream; this charge is hardly credible, for the compound discharged is converted to metallic mercury in the body, and it would then have to be methylated inside the fish to account for the results (the method of analysis again is questioned!). Likewise a plant discharging soluble inorganic and insoluble metallic mercury was blamed for large increases in methyl mercury in fish downstream. It is difficult to believe that the conditions for methylating mercury in bottom muds exist in flowing streams below the necessary dams of mills and plants; the mercury must have come from the water directly without bacterial action in mud.

Curious discrepancies have been found in the concentrations of mercury in different types of fish from the same river or lake. The concentrations may be dependent on size or food habits. The carnivores such as pike and pickerel have the highest levels, and the bottom feeders, such as whitefish and suckers, the lowest. Apparently the mercury in bottom mud sucked in with food does not get into the flesh of the fish, if bottom-mud mercury enters the food chain at all.

The late Senator Winston Prouty wrote:

Consider an area like Silver Lake. It is located in the Green Mountain National Forest, atop a mountain and accessible only by four-wheel drive vehicles or pack trip. It has no industry whatever—but it does have mercury contamination. When Joe's Pond in Danville, Vermont produces fish contaminated eight times beyond the level set by the FDA, we are warned that the issue is far more complex than we had supposed. Vermont is not a heavily industrialized state, and has none of the chemical industries identified as the prime polluters. The orchards discontinued the use of mercury years ago. So did most milk testing plants and paper processors. So where does the mercury come from? [Wallace et al. 1971]

Table VI-3. Effects of Industrial Discharges of Mercury

Mercury in fish and in the food chain from discharges into streams by industrial processes.

Organism	Upstream (ppm)	Downstream (ppm)	Remarks
Fountain moss	0.08	3.70	Paper mill. Effect lasted 12
Water lily	0.02	0.52	miles downstream in all
Leech	0.02	2.35–4.40	organisms.
Isopod	0.06	1.90	
Caddis larva	0.05	5.6–17.0	
Stone-fly larva	0.07	2.4	
Alderfly larva	0.05	4.8–5.5	
Perch	0.18–0.70	1.91–3.02	Paper mill discharging phenylmercuric
Pike	0.55	3.13–3.48	acetate. Said to have 86–100% methyl mercury upstream and downstream.
Eelpout	0.35–0.70	—	Chlor-alkali plant discharging inorganic
Perch	0.20–0.25	0.83–2.48	and metallic mercury. Said to have
Pike	0.11–0.64	1.81	79–100% methyl mercury upstream
Pike perch	0.42–0.66	2.05–2.39	and downstream.
Whitefish	0.05–0.16	1.06–1.40	
Walleye pike	0.11–0.18	1.4–3.6	Lake St. Clair, Mich., and St. Clair
Northern pike	0.21–0.44	0.6	River. "Upstream" values from
Bass	0.07	0.5–0.8	Canadian lakes for comparison.
Coho salmon	—	0.2–1.0	
Others	0.07–0.25	0.1–0.9	
Trout	0.05–4.0	—	Vermont lakes, uncontaminated.

SOURCES: U.S., Dept. of Interior, Geological Survey Professional Paper 713; and Wallace et al. 1971.

The problem is not complex, Senator. The mercury is natural, normal, background mercury untouched by human hands.

Mercury is about five times as toxic as lead, about as toxic as cadmium and antimony, and much less toxic than beryllium. Some forms of metals are more toxic than others. This difference can be due partly to solubility; mercurous chloride, or calomel, is insoluble and is a non-toxic purge, whereas mercuric chloride, or corrosive sublimate, is soluble and when used as a poison by would-be suicides, damages the intestines and the kidneys. Mercurochrome, an antiseptic dye, and mercurial diuretics to increase urine flow can be injected into man's veins with little or no harm.

When a metal is "methylated" by combining with a "methyl group" (such as makes up wood alcohol) it becomes much more toxic. It can also be "ethylated." Tetraethyl lead, the gasoline additive, is a hundred or more times as toxic as lead; so is tetraethyl tin highly toxic, whereas tin is not. These compounds are insoluble in water but soluble in fat, and go into nerve tissue, which is very fatty stuff. So is it with mercury. Methyl mercury is 50 times as toxic as mercury, and it stays in fat and nerve tissue—brain—fourteen times as long. It is very slightly soluble in water. There are many organic compounds of mercury, each with differing toxicities—most are little more toxic than inorganic mercury—but the most toxic is methyl mercury. Therefore, when we speak of mercury in foods, we should specify what kind. Methyl and ethyl mercury can permanently damage the brain when the level in food is 20 to 40 ppm.

Methyl and ethyl mercury compounds have been used for 20 years or more as dressings for seed to prevent mold. They are also added to paint for the same purpose. No one has been reported to be injured from handling them under precautions.

An inadvertent experiment on the effects of large exposures to methyl mercury occurred in Minimata Bay, Japan. A vinyl chloride and acetaldehyde plant making plastics began discharging waste catalysts of methylmercuric sulfide and methylmercuric chloride into the Minimata River on Kyushu Island in 1946 and 1948. The alkyl mercury was taken up by fish and shellfish, which were eaten by local fishermen. A similar situation later occurred in 1951 at Niigata, in the Agano River. These seafoods contained 20 to 40 ppm methyl mercury.

Up to 1971, 184 cases of poisoning were found, the first in 1952. To date, 54 persons have died. In Minimata there were 78 adult cases with 35 deaths, 31 infantile cases with 10 deaths, and 26 fetal cases born with the disease with 3 deaths. The central nervous system was involved in all cases, with permanent degeneration of the brain. Cats, dogs, pigs, crows, and sea birds died. Rabbits, horses, and cows did not. Crabs, fish, and shellfish died.

Of interest are the fetal cases. Their mothers did not exhibit the disease, indicating that methyl mercury is more toxic to the growing fetus than to the adult. All the babies showed signs of cerebral palsy, 7 had small heads, and most of them were severely retarded mentally.

Recognition of this outbreak of poisoning led to the discovery of a number of cases of acute methyl mercury poisoning from the ingestion of seed wheat treated against spoilage with those mercury compounds. During times of food scarcity, treated seeds used as food have caused many deaths in Guatemala, Iraq, and Pakistan; in Alamogordo, New Mexico three people became probably permanently ill from eating a pig fed treated wheat seeds. Prohibition of dumping methyl mercury into lakes and rivers and of using it as a fungistat in seeds and paints should prevent this form of poisoning. These unfortunate episodes proved the toxicity of methyl mercury.

These serious incidents are local problems. The label on mercury-treated seed wheat reads "Warning: Do not use for feed. Contains Mercury." The label on a bottle of bichloride of mercury says: "Warning: Deadly Poison. Keep out of reach of children." The mud at the bottom of Minimata Bay contained elevated levels of mercury, but the sea bottom just outside had background levels, indicating that mercury does not travel far in salt water. No person from other areas of Japan was poisoned by fish.

The march of events proceeded relentlessly, like a fire catching dry grass. The word came to Sweden, where fish from lakes contaminated by dumped mercury were found to contain considerable mercury— at least some fish. In one lake northern pike had 5.8 ppm, whitefish 3.1 ppm, and bottom flounder 0.3 ppm. Pike are fish-eaters. Why the flounder had normal amounts went unexplained, for mercury should be on the bottom. In another lake, these three types of fish had 1.2, 0.6, and 0.05 ppm, respectively. Why the difference in the three kinds of fish?

The Japanese fed cats and rats the polluted shellfish having 20 to 40 ppm methyl mercury and reproduced methyl mercury poisoning. No one did this with Swedish fish or American fish or Canadian fish.

In 1967 a Swedish scientist, Gunnel Westöö, reported that 80 to 90 percent of the mercury in fish, meat, eggs, and liver was methyl mercury. If true this means one of three things: (a) Sweden was polluted with methyl mercury from seeds and dumping; (b) mammals, fish, and birds methylate mercury; or (c) mercury is generally methylated in the environment.

Her findings were confirmed by Swedish and American workers, who concluded that mercurial wastes from industrial effluents were largely converted to dimethyl mercury by bacteria. The hypothesis then went on to say that under acid conditions, as in stagnant muds, dimethyl mercury released methyl mercury, which then entered the food chain. Dimethyl mercury is nowhere near as toxic as methyl mercury.

Toxification of an element is a rare phenomenon in Nature, for it means that the organisms which make it toxic die. It is an antievolutionary idea. Some bacteria, yeasts, fungi, and molds can reduce arsenic to toxic arsene gas, which escapes. But it is unusual for living things to do so; most of Nature's efforts are to detoxify. However, in the laboratory in an aquarium at room temperature some Swedish scientists found some bacteria in mud which would methylate mercury. Lake-bottom mud in the laboratory, however, was only half as efficient. (Not many lake bottoms are as warm as 70 degrees!)

The picture was now clear, with a few gaps which were disregarded. The argument ran: Mercury is dumped in the water, converted to methyl mercury, gets into fish, people eat fish or fish-eaters, and people get sick. (This process was true for methyl mercury in Minimata Bay, but one step was omitted. Methyl mercury was dumped there.) The oceans, now polluted, are converting mercury to methyl mercury at a great rate and all fish have it.

The flaws in this argument are obvious. There is much more natural mercury in the sea than has ever been dumped by man, nearly 3 billion tons in solution. Much more mercury has entered the sea from weathering in the last 100 years to be deposited on the bottom than has been dumped by man, over 350,000 tons. If mercury is converted to methyl mercury on the bottom, this process has been going on for eons, and man's small addition has made little difference. Methyl mercury is quite insoluble, and it is doubtful that it exists in seawater as such. If it does, all the mercury in seawater would have to be methyl mercury in order for it to get into fish. The only known cases of methyl mercury poisoning in humans are from two local areas in Japan, where methyl mercury was dumped. Fish don't cause methyl mercury poisoning when they come from places where no methyl mercury has been added. If fish did cause it, we should expect an epidemic among our sea mammals—dolphin, porpoises, sea otters, seals, whales—some of which are very smart animals.

The ban on game fish rapidly followed the discovery that game fish in Lake St. Clair, Michigan had up to 7.0 ppm mercury and in Lake Champlain, Vermont, up to 2.5 ppm. This news was called a "national catastrophe" by a U.S. Senator. Was it? It began the Great Mercury Scare.

Tuna, 139 of 30,000 batches, was seized as having up to 1.12 ppm mercury. Swordfish was banned entirely as unsafe to eat, and a 12 million dollar industry died. Millions of dollars in fishing licenses were lost as revenue to states. These results were a national catastrophe, brought on by a bureaucratic decision that more than 0.5 ppm mercury in fish was unsafe to eat. The Swedes set their limit at 1.0 ppm. Both countries neglected to do their homework and jumped to conclusions. The 0.5 ppm limit is correct for methyl mercury, but not for other forms which are much less toxic.

It was some time before a few rays of sanity began to penetrate this obfuscation. At the Trace Element Laboratory I had a chance to review the data on Vermont game fish, which have no obvious sources of mercury pollution. Some 37 percent failed to make the FDA grade of 0.5 ppm, and 8.3 percent failed the Swedish one of 1.0 ppm. The Vermont Fish and Game Department caught us a beautiful 30-inch, 10-lb brown trout from Harriman Reservoir, a totally unpolluted area. It had 4.0 ppm mercury, proving our contention that the bigger and older the fish the more mercury it has. (We are saving this fish to eat, when we are through with it.) A scientist who studied coho salmon in one of the Finger Lakes of New York found that the larger and older the fish the more mercury—but it was said to be methyl mercury according to the one method of analysis available. Someone measured mercury in fish preserved since 1927—much mercury. Then prehistoric fish—much mercury. Some of us remembered that J. L. Proust in 1799 found mercury in seawater, and that A. Stock, in 1934 to 1940, showed that all water, foods, and soils, and all mammalian bones, kidneys, lungs, and muscles contained mercury. Not much homework done by the FDA, who set the limit at first at 0.5 ppm, as it was the least concentration detectable by the older method, *not* wanting any mercury in human foods—although the Good Lord put it there in the beginning. (No one bought that idea.)

A list of the natural material exceeding the 0.5 ppm limit of mercury is long: 4 of 11 igneous rocks from the U.S., 0.64 to 1.48 ppm; 33 Russian igneous rocks, 17.6 ppm average, with up to 500 ppm; 3 porphyries, 0.7 to 80 ppm. Limestones up to 10 ppm, sandstones up to 11 ppm, shales up to 60 ppm, sedimentary rocks up to 4 ppm, red clays from sea bottoms 1.8 to 2.0 ppm, mineralized soils 2.5 to 10 ppm, certain coals 1.1 to 46 ppm, and petroleum 1.0 to 500 ppm (Table VI-2). Although federal limits for drinking water are 5 parts per billion, some rivers have 4.2 ppb and groundwaters and springs up to 80 ppb. Human hair, kidneys, lungs, and brain and probably oysters and clams exceed the FDA limits, and people who eat fish or handle mercury accumulate up to 30 ppm in hair. Is all of this methyl mercury? Incredible!

A Swedish scientist discovered that mercury could be rapidly converted to methyl mercury in the test tube by mixing it with methylated Vitamin B_{12}, the anti-anemia vitamin. No bugs were necessary. Actually if you mix acetic acid (found in vinegar) with mercuric chloride and expose it to light you will get methyl mercury. Our laboratory found that 1.0 ppm methyl mercury improved the growth of young mice, but 5 ppm had the opposite, or toxic, effect. Another Swedish scientist, whose findings were not publicized, found only traces of methyl mercury in sea fish, although the FDA says swordfish are full of it. (I just don't believe it.)

Whenever data don't fit the vast majority of facts, one should scrutinize the method by which the data were obtained. The Swedish method of measuring methyl mercury depends on digesting the sample of food (fish, meat, eggs, etc.) in strong acid. Acid breaks down proteins. All proteins are made of amino acids, some of which contain methyl groups. If these are loosened, they can latch on to the mercury present, making methyl mercury. If one can do it with vinegar and acid, it is possible, but not yet proven, that the method of analysis makes methyl mercury when there was none before. Or it may be that fish—and we—contain dimethyl mercury in small amounts, and that the acid treatment releases methyl mercury. Thus, the whole mercury scare could be a gigantic scientific misconception, inadvertent, to be sure, but still false.

One of the silliest projects of this whole fiasco was the result of some government measurements of Alaskan Aleut Indian hair. The luckless Aleuts had 10 times as much mercury as the FDA allows in fish —actually less than many Americans, Englishmen, and Welshmen have, but the homework was not done. Scientists traced the source to seal blubber. Seals eat fish, Aleuts eat seals, the oceans are polluted, fish are therefore polluted, seals are therefore polluted, the Aleuts are therefore polluted and are suffering from mercury poisoning. Maybe the far-ranging seals get their mercury from fish off California, polluted by mercury dumped into rivers during the Gold Rush in 1849. (Although I doubt it. They get the mercury from any fish.) It all hangs together—except for the fact that the oceans are not polluted and the Aleuts are not poisoned; at least no Aleut has complained of poisoning to our knowledge.

The dust is settling. It is unlikely that much mercury is methylated in the ocean bottoms. It is unlikely that mercury is methylated in lake bottoms to any significant extent. It is clear that methyl mercury is toxic to man. It is equally clear that natural mercury in foods is not toxic to man (no one has produced methyl mercury poisoning with swordfish).

We must avoid foods contaminated by methyl mercury and forget about foods with natural mercury. Dumping of methyl mercury must stop, and it is just as well to stop dumping mercury into lakes. We can eat our game fish without harm or alarm. It is time for the FDA to backtrack, as they have with other pollutants, and bring back game fish and swordfish to our tables.

Fish are an excellent source of protein. Poor people can catch fish all year round for nothing. In Vermont, there are many poor people who catch and depend on fish for food. With characteristic Vermont skepticism, they have listened to the alarms and gone on fishing. Our advice to fisherman is to do the same—except in waters polluted with methyl mercury. Vermont has lifted the ban on swordfish. So has New Hampshire. Let other states do the same.

VII

CADMIUM, THE DRAGON'S TEETH

Doctors concerned with occupational diseases have long known that cadmium is poisonous. Pharmacologists, also, have known that small doses lead to bizarre effects in animals. At the turn of the century, cadmium was a rather rare metal nearly always associated with zinc, little used industrially, a metallic curiosity with a bluish-white lustre quite impervious to corrosion, like tin.

World War I brought enormous demands for tinned food and a shortage of tin, so that a substitute was needed. Cadmium was ideal though expensive, but its possible toxicity in foods required investigation. In the early twenties, a deluge of literature appeared showing that cadmium was not suitable for coating food containers; small amounts dissolved by acid foods and drinks made animals and people acutely ill.

The reader may remember the old Greek myth about Europa, daughter of the King of Phoenicia, who was abducted by a traveling bull and from whom Europe gets it name. Cadmus, her brother, was sent by his father to find her. Cadmus consulted the Delphic oracle and was commanded to give up the search, for Zeus had the lady in Crete for his own amorous purposes. Instead, he was told to follow a certain traveling cow and to build a town where the cow sank down exhausted. He chased the cow through two countries and built the citadel of Thebes where she sank. Cadmus then sent some people to fetch water from the well of Ares, which was guarded by an unfriendly dragon. The dragon slew them, as that was his job. Cadmus killed the dragon, and on the advice of Athena, ploughed a field and sowed the dragon's teeth. A regiment of fully armed fighting men sprang from the ground, charged, and turned on each other until all but five were killed. These five were the first Thebans, the Sparti, or "sown-men."

Cadmium has somewhat the same properties as the dragon's teeth. It looks innocuous but it has a vast potential to poison. Unlike the sown-men, it is not recognizable as an enemy, acting subtly and under-cover, mimicking diseases in man for which other causes have been proposed, accumulating in the body slowly until the threshold of resistance is overcome, then striking. This subtle property was not recognized until recently.

With the loss of Malayan tin in World War II, cadmium was again considered as a substitute for tin cans. As a result, another spate of papers appeared in the middle forties again showing its animal toxicity. The memory of Science is often short.

In spite of its toxicity, cadmium was used to coat ice trays in electric refrigerators. Some people like to make lime or lemon sherbets by freezing the mixes in ice trays. The acid will dissolve some cadmium, and in the mid-thirties there were several cases of acute cadmium poisoning with a few deaths from this source. Dr. Thomas Arthur Gonzales, then assistant medical examiner of New York City, tracked it down, and cadmium-coated ice trays were banned—in New York. (But not everywhere. In the late fifties, I discovered four such trays in my General Electric refrigerator—and banned them.)

Dr. Gonzales, being a thorough scientist, analyzed the organs of his cadmium victims by the method then available. He found the kidneys loaded with cadmium—but so were the kidneys of people dying of other causes. He gave up the study, and it was not until 1953 that Dr. Isabel H. Tipton found much—but varying—amounts of cadmium in kidneys of all the adult Americans she examined, but little or none in babies' kidneys, thus starting us on our worldwide search for cadmium and its sources.

Cadmium became well established as an anticorrosive plating on metal parts during World War II. It was virtually essential for aircraft exposed to salt spray, and hardened aluminum nuts, bolts, cylinders, small parts, and valves showed the characteristic pale yellow color of cadmium plating. In 1944 I took a hundred gravity-operated valves, part of automatic equipment for pilots' anti-blackout suits, on an aircraft carrier far into the Pacific Ocean for three weeks. They lasted two weeks under operating conditions. When similar valves were anodized with cadmium, not a single one out of some fifty thousand failed from corrosion. Nor has one failed since.

Thus, cadmium is replacing zinc as a plating on metal. Some 16,000 tons are used yearly and consumption rises every year. In New York City alone, there are 152 small companies engaged in cadmium plating. Metal parts are put into an electrolytic bath containing cadmium salts, plated, and then removed and washed off with fresh water. The drippings go down the drain. The cadmium enters the city's sewage treatment plants, collects, and in time poisons the bacteria digesting sewage and garbage. The treatment plant then has to shut down for several months, in the interim discharging raw sewage into the rivers and the harbor. Not only is raw sewage an unpleasant pollutant, but the cadmium precipitates in salt water to the bottom mud, where it can enter the food chain. Just a little cadmium, a half to one part per million in water, is toxic to most bacteria tested.

Two lawsuits have been brought by the U.S. Department of Justice. One is against the City of New York, the mayor, his environmental protection administrator, his water commissioner, and two electroplaters as representatives of a class of about 200 firms for polluting federal waterways with cadmium, other heavy metals, and toxic substances. The second is against the State of New Jersey and its many towns and cities discharging into the Hudson River and New York Harbor. These suits demand compliance with city, state, and federal regulations on pollution of the aquatic environment with antimony, arsenic, barium, boron, bromine, cadmium, chromium, copper, fluoride, gold, iron, lead, manganese, magnesium, mercury, molybdenum, nickel, rhodium, selenium, silver, thallium, titanium, tungsten, vanadium, and zinc (see Table VII-1).

If the Department of Justice wins, and discharge of cadmium into sewers virtually ceases, it will still be too late for the fish in the Hudson River. The Trace Element Laboratory and the Environmental Protection Agency have evidence that over half the fish in the Hudson River are unsafe to eat regularly because of contamination with cadmium.

One cannot predict how much cadmium is in fish by measuring it in water. The Hudson River has very little cadmium dissolved in it, a few (3–6) parts per billion, but its fish have a good deal. Four Alabama rivers had 6, 12, 65, and 90 parts per billion cadmium, re-

Table VII-1. Metals Found in 720 Rivers and Lakes of the U.S.

Metal	Allowable limits (ppb)	Concentrations within allowable limits		Concentrations above allowable limits	
		Bodies of water (%)	Range (ppb)	Bodies of water (no.)	Range (ppb)
Mercury	5	7	0.1–4.3	0	—
Cadmium	10	42	1–10	31	12–130
Lead	50	63	1–50	2	55, 890
Arsenic	50	21	10–50	9	140–1,100
Chromium (VI)	50	1.5	6–50	0	—
Cobalt	—	37	1–5	—	—
Zinc	5,000	100	10–50	2	>5,000

SOURCE: U.S., Dept. of Interior, Geological Survey Circular 643.

spectively; about the same amount of cadmium, very little, was found in fish from these rivers (Table VII-2). Compare these Alabama fish with Hudson River fish and the difference becomes obvious. The Hudson River cadmium is in the mud and the food chain; in the Alabama rivers cadmium is in the water and in the mud at the source but not downstream. It is cadmium in mud, not in water, that we must worry about.

In 1972, a lawsuit by the Department of Justice against the Marathon Battery Company and Sonotone Corporation was won by consent of the defendants. These companies made cadmium-nickel batteries for warplanes, hearing aids, power tools, and electric shavers and were discharging wastes into Foundry Cove on the Hudson River. Mr. David M. Seymour, a worker for the Audubon Society, and Robert H. Boyle, a free-lance writer on sport fishing, were walking by the cove and noticed that the mud had a greenish grey color, like pea soup. They sent us some of the mud and Alexis P. Nason, our laboratory analyst, found that it contained over 16% cadmium and 22% nickel. There were an estimated 25 tons of cadmium and 32 tons of nickel in the cove, which at $2.00 and 80 cents a pound, respectively, were worth about $150,000. They asked me what to do. "Mine it!" I answered. But they didn't. They began catching fish and sending them to us until Mr. Nason was piled high with work.

About then, Senator Philip S. Hart entered the picture, and his people persuaded us to analyze the 44 Alabama fish listed in the table. Our freezers were full of fish. Mr. Nason analyzed them and said, "No more. The law of diminishing returns is now operative." We had learned most of what we wanted—and needed—to know: cadmium in river water was not necessarily reflected in its fish.

The Department of Justice then entered the fray, and persuaded the EPA to do the analyses. Which they did, on everything. It added little to the case, except to show that most Hudson River fish were contaminated, as was the food chain (which we knew).

In December 1971 our laboratory was turned into a miniature courtroom, with lawyers for the prosecution and defense, federal court reporter, oaths, and all the works but a judge. Transcription of my testimony was three inches thick, and so garbled that I never got around to correcting it. But it was enough. The next June the EPA announced the verdict; we, who helped discover the stuff, did not even get honorable mention. But the defendants removed most of the cadmium and nickel from Foundry Cove, thus preventing a hundred years of further contamination of the Hudson River. Fortunately, there is yet no cadmium lobby or propaganda institute.

The Hudson River is a local and regional problem. So is the Jintzu River in Japan. A large lead and zinc smelter was discharging its tailings where they were washed by river and rainwater. The river was used for irrigation and drinking by people living downstream. Cadmium and lead entered rice and wheat crops, fish, and people. They accumulated. After many years, a disease named *itai-itai*—ouch-ouch —appeared in women past the menopause. Calcium was lost from bones, and they fractured easily. Kidneys were badly damaged. Deformities were severe. Many died. The victims' organs were loaded with cadmium and lead. Even their bones had much cadmium, and cadmium does not usually enter bone. It will probably take a hundred years for that soil and river bottom to cleanse itself of lead and cadmium.

All that happened from a zinc smelter. Cadmium is constantly present in zinc, even the purest. The U.S. Bureau of Standards' standard zinc slab has 0.53% cadmium. The zinc used for galvanizing iron— tin roofs, pails, water storage tanks, iron pipes, gutters, water-softening tanks, maple-sap buckets, cauldrons, barbed wire, chicken fences, wire fences, animal cages, nails, and a host of other articles of iron or steel —is far from pure, generally being the cheapest grade, which contains even more cadmium. Impure zinc can contain up to 2%.

Table VII-2. Cadmium Concentrations in Fish

Cadmium in flesh of fish and shellfish from contaminated and uncontaminated waters, average values, wet weight.

Place and kind of fish	Cadmium in water (ppb)	No. fish	Cadmium in fish Flesh (ppm)	Kidney (ppm)	Other (ppm)	Remarks
Shell Bayou, Ala.	—					
Bass and bluegill		4	0.02	—	—	
Black Warrior						
River, Ala.	6					
Bass and bluegill		8	0.03	—	—	
Drumfish		1	0.03	—	—	
Cahaba River, Ala.	12					
Bass and bluegill		2	0.01	—	—	
Catfish		2	0.02	—	—	
Sunfish		5	0.04	—	—	
Mobile River, Ala.	65					
Bass and bluegill		5	0.02	—	—	
Redear		4	0.02	—	—	
Catfish		4	0.02	—	—	
Tennessee River, Ala.	90					
Bass and bluegill		8	0.08	—	—	
Redear		1	<0.01	—	—	
Foundry Cove,	5–26					Mud content,
Hudson River						16.2% cadmium
Dace, minnows		5	9.1	—	11.8	Minnows, whole and innards
Bass, minnows		16	12.3	—	53.5	Minnows, whole and innards
Turtle eggs		10	0.005	—	—	
Hudson River,						
indeterminate	3–5					
Carp		4	0.2–1.2	1.8–39.7	—	
Golden shiner		4	0.2–68.9	6.4–16.29	—	1 contaminated
Perch, white		5	0.3–304	3–15.7	0.4	1 contaminated
Turtle eggs		10	0.05	—	—	

SOURCES: Analyses on Foundry Cove, Montrose, and other fish by Trace Element Laboratory. Analyses on Hudson River, indeterminate, by Environmental Protection Agency. Analyses of oysters and clams by Food and Drug Administration, Shellfish Sanitation Branch.

Table VII-2 (cont.)

| Place and kind of fish | Cadmium in water (ppb) | No. fish | Cadmium in fish | | | Remarks |
			Flesh (ppm)	Kidney (ppm)	Other (ppm)	
Montrose, N.Y.,						
Hudson River	3–5					
Carp, 20 lb		1	0.4–0.67	20.44	1.2–12.3	7 tissues 0.13–0.83
Sunfish		1	0.24	—	—	
Herring		1	0.16	—	—	
Perch		1	0.25	—	—	
Striped bass		1	0.25			
Miramichi, N.B.	—					
Atlantic salmon		1	—	—	0.26	Liver
North Atlantic Ocean	0.03	—	—	—	—	
Halibut		1	0.05	—	—	Sample from supermarket
Haddock		1	0.09	—	—	Sample from supermarket
Unknown						
Fish flour, dry		—	1.56	—	—	Food, whole fish ground
California mussels, dry		—	4.93	—	—	Beds may be contaminated.
Atlantic Ocean	0.03					
Oysters: Average		—	3.1*	—	—	Beds may be contaminated.
Maximum		—	7.8*	—	—	Beds may be contaminated.
Clams, hard:						
Average		—	0.19	—	—	
Maximum		—	0.73	—	—	
Clams, soft:						
Average		—	0.27	—	—	
Maximum		—	0.9	—	—	
Gulf of Mexico	0.03	—	—	—	—	
Oysters: Average		—	0.88	—	—	
Maximum		—	2.6	—	—	

* The high values in oysters are accompanied by 1,428–4,120 ppm zinc, and are probably not harmful. Any fish, however, with more than 0.25 ppm cadmium should be eaten with caution.

Whenever a slightly acid liquid comes in contact for a time with galvanized metal it dissolves some zinc and cadmium. Rainwater is slightly acid from the dissolved carbon dioxide in the air. Falling on a tin roof, collected by a gutter, stored in a galvanized iron cistern, rainwater will contain zinc and cadmium, for it is slightly corrosive. Soft water is also corrosive, and when it stands overnight in galvanized iron pipes, it dissolves zinc and cadmium. When you draw water for your morning coffee before flushing the pipes by running the water, the cadmium is in the coffee. Most of the galvanized iron plumbing in the old houses of our town has been replaced with copper, for our soft acid water has corroded the pipes. First the zinc and cadmium plating goes, then the iron, which makes the water rusty, and then a leak springs and there is a hurried call to the plumber. Soft water also corrodes copper from pipes—bluish green stains on the toilet bowl and the bath tub are characteristic. Hard water is usually not corrosive. Hot water is much more corrosive than cold.

A number of cases of so-called zinc poisoning have been caused by drinking lemonade which has stood for several hours in galvanized pails, cauldrons, or washtubs. Actually they were due to cadmium poisoning, for zinc is not poisonous except in huge doses. Zinc is necessary for all living things, and plants and animals grow poorly or die when there is not enough zinc in their environments. In zinc-deficient soil, a half dozen galvanized nails driven into the trunk of a fruit tree make the difference between health and disease; a string of barbed wire can make healthy crops grow near the fence. Zinc-deficient chickens can get enough zinc by pecking at the wire on their cages or pens. So can rats by licking galvanized wire. But they also get the cadmium.

The human body contains about 2.2 grams of zinc, and there are mechanisms which keep this amount constant throughout life, unless the diet is low in zinc. The body contains a variable amount of cadmium, normally 30 milligrams in Western society, as little as 10 milligrams or less in certain African nations, as much as 50 or 60 milligrams in some people, especially the Japanese, who have more than the Europeans. Zinc does not accumulate. Cadmium, once in the body, stays, probably for life, in the kidneys, the liver, and the blood vessels. As little as 2 micrograms daily absorbed and retained results in a body burden of 30 milligrams in 40 years. Cadmium displaces zinc but does not act beneficially like zinc; quite the opposite. The human kidney contains 8 to 10 times as much cadmium as do the kidneys of any other mammal, except those pets exposed as we are.

Where does the cadmium come from? The air of some cities contains cadmium from fumes spewed out by zinc smelters and refiners and copper smelters. More than 2,000,000 lb (about 1,000,000 kg) are released into the air space of U. S. cities annually from this source alone. An additional 2,300,000 lb went up the chimney in 1968 from the recovery of scrap metal, and another 300,000 lb from incinerators. World consumption is 31 million lb, of which the U. S. uses 12 to 13 million lb. When we swallow cadmium, only a small amount is absorbed into the body from the gut, perhaps 10%, most of which is excreted in the urine. When we breathe cadmium, we retain about half of it, absorbing it from the lungs. A pack of cigarettes contains 16–24 micrograms of cadmium, and a smoker can contaminate a whole roomful of people.

Rubber tires, plastics, pigments, plated ware, alloys, insecticides, and solders are some of the things containing cadmium. Some foods have a lot of cadmium, relatively: oysters, foods contaminated in the processing, some instant coffees and teas, some canned foods, kidneys of pigs given cadmium as a worm killer, gelatin and fish dried on chicken wire, some cola drinks. We used to get it from dental fillings, but we don't any more. Pigments can be a source, for cadmium yellow and cadmium red are fast colors; some French lipsticks have it. A necklace of candy "LUV" beads made in Hongkong was colored with cadmium; it made one little girl sick for over a year and poisoned her brother for a short time.

Cadmium is so ubiquitous in our civilization that it is very difficult to avoid it. Our laboratory developed a diet for rats and mice which is very low in cadmium, so low in fact, that it does not accumulate in their bodies over a lifetime. But we have to keep our animals in a metal-free laboratory on a remote Vermont hilltop, give them absolutely pure water, and take extensive precautions to avoid contamination. All commercial diets contain too much cadmium for our use.

What is the price we pay for environmental cadmium? High blood pressure is one. And a major one. The earliest sign of subtle cadmium toxicity is elevation of the blood pressure (Table VII-3). The usual upper limit of normal for rats' blood pressure is about 140 millimeters of mercury. With only a trace of cadmium in the diet and with only a trace in the kidneys and liver, it was a bit over 80. When rats were

Table VII-3. Effects of Cadmium in Food and Water

Blood pressures and accumulations of cadmium in the kidneys and livers of rats fed diets of differing cadmium concentrations.

Age, months	Cadmium intake			P
	0.07 ppm	0.68 ppm	5.1 ppm	
Systolic blood pressure, mm of mercury				
Females				
3	85 ± 2.2	109 ± 3.2	—	<0.001
4	87 ± 4.4	110 ± 3.7	—	<0.001
5	81 ± 2.2	112 ± 6.1	—	<0.001
7	—	115 ± 4.0	—	<0.001
12	84 ± 5.8	—	211 ± 8.3	<0.001
13	82 ± 3.4	—	—	—
17	92 ± 4.9	—	182 ± 12.6	<0.001
24	84 ± 3.8	—	205 ± 10.9	<0.001
30	99 ± 4.2	—	229 ± 12.9	<0.001
Males				
12	106 ± 5.7	—	124 ± 5.6	<0.025
17	94 ± 3.8	—	122 ± 4.5	<0.001
24	79 ± 3.6	—	137 ± 6.2	<0.001
30	93 ± 5.1	—	198 ± 7.9	<0.001
Accumulations				
Renal cadmium, ppm	0.03 ± 0.002	0.76 ± 0.060	54.7 ± 4.82	<0.0001
Hepatic cadmium, ppm	0.03 ± 0.008	0.58 ± 0.160	20.9 ± 3.91	<0.0001

Note: One group of 20 rats was weaned on our low-cadmium diet containing 0.07 ppm cadmium, another group of 20 on a commercial rat chow containing 0.68 ppm. The first two columns show blood pressures, taken while the rats were asleep and warmed. For the first seven months of life pressures were 23–34 mm higher for the rats taking the commercial chow than for those taking our own diet. When rats on our diet were given 5 ppm cadmium in water, blood pressure was 90–130 mm higher in females and 18–105 mm higher in males. The last column shows the probability that the difference is due to chance—in most cases less than one in a thousand. Rats on our diet did not accumulate cadmium, but those on the commercial diet did and especially when given it in water; some human beings have as much accumulated cadmium as those rats.

fed a commercial diet containing cadmium and some deposited in their kidneys, their blood pressure was about 110. Not high, but higher. When cadmium was given in drinking water and there was much in the kidneys, liver, and blood vessels, the rats had high blood pressure, with large hearts, thickening of the small arteries of their kidneys, and, in some cases, heart attacks and hemorrhages. They also showed hardening of the arteries.

High blood pressure from cadmium has now been produced in rats, rabbits, and dogs. It is not severe, usually, but is the spitting image in all respects of the kind 23 million Americans have, which promotes heart attacks and strokes. When we look at human kidneys, we find that people who died with high blood pressure had either more cadmium or less zinc in their kidneys than did people who died of other causes. Cadmium had displaced zinc and slightly poisoned some zinc system that controlled blood pressure.

We played a trick on Nature. The laboratory developed a drug which contained zinc but which would chelate—bind, or grab— cadmium where it met cadmium and drop zinc in its place. High blood pressure in rats was cured very quickly; some of the cadmium was removed and some zinc put back. In people, this drug has lowered blood pressure for long periods of time, removing a little cadmium from the right places, probably the blood vessels.

In essence, we reproduced a common human disease in animals by using a common toxic metal, cadmium, cured it by removing the metal, found a similar situation in man, and relieved it in the same way. Although there is little more to do except clean up a few minor pieces of the picture puzzle, it will take ten to fifteen years before the medical profession will accept this novel idea, in spite of the fact that our work has been confirmed by others.

There is a curious phenomenon in this story. The right amount of cadmium over a lifetime causes high blood pressure. Too much cadmium does not. When cadmium accumulates beyond the subtle poisoning stage, the kidneys and the liver are damaged and blood pressure falls. The same effect occurs when cadmium is injected. A little raises blood pressure; a lot lowers it. The Japanese with ouch-ouch disease did not have a high incidence of high blood pressure— they were too sick. Workers in cadmium battery factories, who breathe a great deal of cadmium into their lungs, do not necessarily have high blood pressure—they are too poisoned. There are many examples of this kind of effect in medicine—the right dose of digitalis improves a failing circulation, but too much increases failure. Nicotine in small doses stimulates nerves; in larger doses it paralyzes.

Emphysema of the lung is a nasty disease. The little air sacs of the lungs rupture, making larger ones. Breathing becomes labored. High blood pressure in the circulation of the lung appears, straining the heart and leading to heart failure. Emphysema patients also have more cadmium in their kidneys and livers than do well people. Cigarette smoke contains cadmium, and it is tempting to guess that cadmium absorbed directly by the lungs initiates high blood pressure there. Emphysema is common in cadmium workers. Our drug might be of value in treating emphysema.

When fed to breeding rats and mice, cadmium causes severe congenital abnormalities, to such extent that the strain dies out. When injected into pregnant rats, it produces toxemia of pregnancy, and into pregnant hamsters, congenital abnormalities in the offspring. Injection is not a fair way to test anything we eat, drink, or breathe, but it serves to show bizarre toxic effects.

Did anyone ever think of injecting waste water from washing machines and dirty dishes into pregnant women? Of course not. Yet that was what the Environmental Protection Agency and the Public Health Service implied when they banned NTA last year.

NTA—nitrilotriacetic acid—like the polyphosphate detergents, is a chelating agent, binding metals. That is its virtue. NTA was allowed as a good substitute for phosphates, which are highly nutritious to plants and cause overgrowth of algae in stagnant lakes, choking off fish life. Some 100 to 125 million tons of NTA were made annually as a substitute. NTA is rapidly biodegradable by oxygen-dependent bacteria, a distinct advantage over phosphates.

Some experiments were done at the National Institute of Environmental Health Sciences—rather rapid ones, I suspect from the data— in which NTA was *injected* into pregnant rats. No effect. Cadmium and methyl mercury were also injected. No effect. Combined with NTA, cadmium and mercury were both lethal to fetuses and mothers.

The idea is that cadmium and methyl mercury are common water pollutants. NTA gets into the water from drains and combines with methyl mercury and cadmium. Pregnant women drink the water somewhere else and get dead babies. Synergism. The NTA allows the metals to get into the body and pass through the placenta into the fetus.

To anyone with a knowledge of trace metals, there are several glaring flaws in this argument:

1. Although methyl mercury probably occurs in waters where it has been dumped, it has never been demonstrated to exist as such in any water, and it is probably rare in the United States. Regular mercury is found in concentrations of less than 5 parts per billion; such small amounts are harmless, NTA or not (Table VII-1).

2. There is a little extra cadmium in some rivers, but most waters have less than 10 ppb, too little to harm a fetus.

3. No one has demonstrated that NTA by mouth increases the absorption of cadmium or methyl mercury by the intestines. Indications are that it doesn't, for most other metal chelates are poorly absorbed —the molecule is too big.

4. Large doses of cadmium and methyl mercury injected into pregnant rats did nothing, although much smaller doses by mouth are bad both for the young and the mothers, as several people have discovered. There was something fishy about the NTA experiments.

Two Swedish technicians came all the way to the Virgin Islands to consult about NTA. They were worried about the data. On critical examination they did not stand up. Neither did the experiments. The Swedes went home satisfied to continue the use of NTA.

The following year the Public Health Service backtracked quietly and lifted the ban on NTA, which is now clean as a hound's tooth, although slightly tarnished in reputation. So it goes.

One of the subtle effects of this product of the dragon's teeth is that no other measurable function of the body is altered, other than the level of blood pressure and what goes with it. The patient does not know that a function of his kidneys detectable only by sophisticated techniques is changed, nor does he feel the difference in his blood pressure, except under anger or anxiety, when he may flush or his heart may pound. Yet his heart works harder with each beat, day and often night, gradually enlarging; the very small arteries of his kidneys constrict in an attempt to protect the delicate capillaries which filter his blood to make urine, and both show signs of strain; he already has a serious blood vessel disease, arteriosclerosis, and the heightened blood pressure increases its progression and makes it worse, both in his large arteries, which do not matter too much, and in the smaller ones of his heart and brain. The process, a vicious cycle, goes on inexorably, until one day in his middle age, there is a sudden accident. An artery in his heart is narrowed to the point where it supplies insufficient blood for the needs of that part of the heart, and he feels the severe pain of angina pectoris. Or the artery plugs, and he has a heart attack, or coronary occlusion. Or, later in life, the same thing happens to an artery in his brain, and he has a stroke or thrombosis. Or an artery in his brain ruptures, and he has a cerebral hemorrhage. Something like that kills over half the population of our country.

High blood pressure can now be treated with modern drugs and is often reversible, unless it has gone too far. When well treated, its serious consequences, heart failure and hemorrhage, largely disappear. Unfortunately, it is not as well treated as it should be, even though the death rate from hypertensive heart disease is one quarter of what it was 20 years ago. If we could remove the cadmium from blood vessels, replacing it with needed zinc, regular treatment with drugs could become unnecessary, except at long intervals.

The amount of cadmium we take into our bodies seems to depend on the amount of zinc we also absorb; the more zinc, the less cadmium. We can speculate with reasonable assurance that plenty of zinc may displace a little cadmium in the tissues. We know that animals deprived of cadmium have less zinc in their bodies than do animals fed cadmium, which have twice as much, although they seem healthy. Cadmium demands zinc. Today, with the wide use of refined flour, sugar, and fats, we are not getting enough zinc for our needs, especially if cadmium is around in our food, water, and air. Many Americans, especially teenagers and the aged, have measurable zinc deficiencies.

A major breakthrough in the treatment of poor circulation of the legs resulting from narrowing of the arteries is the use of large amounts of zinc by mouth. Pain ceases, normal color returns, exercise tolerance improves, ulcers and gangrene heal, and the affliction is cured. We don't believe that zinc reverses the arteriosclerosis of the vessels—though it may. Rather we speculate that zinc displaces cadmium and reverses the spasm induced by it in minor arteries. Zinc also works in angina pectoris.

When an artery is narrowed at one point, the small arteries it supplies downstream become highly sensitive to material in the blood which may normally constrict them a little. The end result is marked loss of blood flow. Zinc prevents that sensitivity.

Zinc by mouth is also of value in loss of sense of smell, which can be caused by cadmium, but cause and effect have been established only in cadmium workers. Zinc increases healing of wounds, for it is necessary for the growth of cells. I have also seen it improve tolerance for alcohol, and it has been used with some success in cirrhosis of the liver.

The skeptical reader may ask "Haven't we always had high blood pressure, before cadmium was introduced into the environment?" We have, associated with kidney diseases, at least since 1693, but it has been nowhere near as prevalent as it is now. In certain parts of the world, for example, Burundi, Africa, it occurs only when the kidneys are diseased. The moderate and extremely frequent type seen in this country is only found in civilized societies or in places where there is obvious exposure to cadmium in water—certain islands of the West Indies, where rainwater is collected on galvanized roofs and stored in galvanized cisterns, for example.

That hypertension is mainly an environmental disease, although it may have a hereditary background, becomes clear when one examines the whole picture. We do not have space for more than the highlights. The Negro race is supposed to be very susceptible to hypertension; in this country, it is, with three or four times as many deaths as among whites. Hypertension is common in West Africa, but not among Central African Negroes. The rate for Negroes is equal to the white incidence in the Virgin Islands but twice as high as the rate for whites in St. Kitts. In other words, the incidence in Negroes varies from negligible to very high depending on the area. The incidence of hypertensive deaths in white people varies from city to city in the U.S. by as much as four times. From country to country it varies by as much as ten times. It is very high in Japan and very low in Thailand. In 94 cities of the United States, the death rate from heart attacks varies directly according to the corrosiveness of finished municipal water (see Table VII-4); in Japan, the death rate from cerebral hemorrhage varies according to the corrosiveness of river water. Death rates from heart diseases vary according to water quality in Britain, Canada, Sweden, the Netherlands, and South America. It is the pipes, and probably cadmium in the pipes, which are responsible.

There is everywhere a marked difference in death rates from heart disease in backward countries between the lowest class and the lower middle class. The first thing a poor family does when it establishes some measure of economic solidarity is to move into a house with running water. There they begin to have heart attacks. Doctors argue that such people eat more fat, or sugar, or what-not, thus accounting for heart disease. Pipes seem more logical villains.

Table VII-4. The Role of Drinking Water in Deaths from Heart Diseases

Correlations of death rates per 100,000 population from diseases of heart and blood vessels with corrosiveness and hardness of municipal drinking water in 94 largest cities of U.S., 1959–1961.

	White				Nonwhite			
	Age 45–64		Age 65+		Age 45–64		Age 65+	
	Male	Female	Male	Female	Male	Female	Male	Female
Arteriosclerotic Heart Disease								
Mean Death Rate	636	157	2,729	1,548	526	296	1,851	1,298
Coefficient of Correlation (r):								
Index of Corrosion	−0.293	−0.367	−0.424	−0.454	—	−0.201	—	−0.369
Hardness	−0.330	−0.237	−0.237	−0.298	—	—	—	−0.288
Cerebral Hemorrhage								
Mean Death Rate	61	50	506	470	133	173	544	533
Coefficient of Correlation (r):								
Index of Corrosion	—	−0.181	—	—	—	—	—	—
Hardness	−0.278	—	—	—	−0.192	—	—	—
Hypertensive Heart Disease								
Mean Death Rate	32	25	221	258	133	148	441	506
Coefficient of Correlation (r):								
Index of Corrosion	—	−0.312	—	−0.212	—	—	—	—
Hardness	—	−0.222	—	—	—	—	—	—

Note: Statistical significance of the coefficients of correlation are as follows: $r = 0.173$, $P = 0.05$; $r = 0.203$, $P = 0.025$; $r = 0.241$, $P = 0.01$; $r = 0.265$, $P = 0.005$; $r = 0.335$, $P = 0.001$. Negative correlations mean that the death rate became lower as the concentration of the water constituents increased or as corrosiveness decreased. Thirty-five constituents of water were analyzed and statistically treated; corrosiveness was usually the most important factor in arteriosclerotic heart disease, but not in other diseases of the arteries. In this respect, Stitt et al. 1973 have shown that middle-aged men living in soft-water areas of England had higher blood pressures, plasma cholesterol levels, and resting heart rates than a matched group of men living in hard-water towns.

Sources: Schroeder, Kraemer.

What can we do about the situation with this subtle, accumulative poison, a clear and present hazard to health? First, we can prevent its emission into the air. Whenever zinc is burned or melted there will be cadmium. Incineration is a major source: burning of automobile tires, red and yellow plastic bags, plastic products, paints, discarded automobiles, discarded airplanes, and the parts thereof. Cadmium pollution can be abated by prevention of air pollution with zinc (Table VII-5).

Table VII-5. Potential Exposures of Human Beings to Cadmium

Source	Remarks
From Air	
Smelting of zinc, lead, and copper	In chimney exhausts.
Incineration of plastics and pigments	In chimney exhausts. Pigments account for 22% of total consumption.
Incineration of rubber goods and tires	In chimney exhausts.
Wear of rubber tires.	In roadside air. Cadmium is a stabilizer for rubber.
Burning motor oil	In chimney exhausts and motor vehicle exhausts.
Motor vehicle exhausts	Small amount in gasoline, large amount in exhaust system.
Burning cadmium batteries	In smoke over dumps. 3% of total consumption.
Cigarette smoking	Content 20 μg/pack.
Cigarette smoke in rooms	In particulates. Affects nonsmokers.
Melting of used cars and aircraft for scrap	In exhausts. Much cadmium plating per car or airplane.
From Water and Fluids	
Electroplating	In wash water to sewer. 45% of total consumption.
Metal alloys	Solution in water or acid fluids. 7.5% of total consumption.
Fungicides	Golf courses to streams.
Polyvinyl plastics	Used as plasticizer. 15% of total consumption.
Galvanized iron pipes	Soft and acid waters dissolve from zinc coating.
Galvanized iron roofs	Dissolved by rainwater.
Galvanized iron cisterns and tanks	Dissolved by soft water.

Table VII-5 (cont.)

Source	Remarks
Cola drinks	From processing. Usually 10 μg per quart or less.
Instant coffees	From processing.
	From Food
Oysters	Up to 7 ppm, with much zinc.
Some canned and dried fish	From canning, drying, or smoking. Galvanized wire netting?
Pigment in candy	"Luv beads" make children ill.
From plastic wrappings	Absorbed by food.
Plated ice trays	Dissolved by acid sherbets.
Plated roasting pans	Dissolved by roasting fats.
Pigmented pottery	Dissolved by acid foods and juices.
Silver polish	Residue on eating utensils.
Lipstick, imported	Swallowed by women and transvestites.
Pork kidneys	Cadmium used as a vermifuge.
Butter	Probably from galvanized milk cans.
Olive oil	From cans and presses.
Gelatin, dried fish	From galvanized netting for drying.
Many processed meats	From contact in processing machines.
Tin and aluminum cans	Tin cans made of old cars, aluminum from old aircraft.

Second, we can prevent its solution into water—along with lead—by providing municipal waters that are not corrosive. It is not difficult. Almost all municipalities treat their water, but not for corrosiveness.

Third, we can begin to control cadmium entering food and drinks by careful monitoring—by prevention of dumping into rivers and estuaries, by restrictions on its use in food containers—in other words, as we try to control any poison. In Table VII-5 are some examples of foods containing cadmium. Others with more than 1 ppm are smoked kippers, canned anchovies, lamb chops, chicken, olive oil, instant coffees and teas, tea leaves, and caffein-free coffee; some with more than 0.5 ppm are all seafood, meats, wheat gluten (cadmium goes with the gluten in grains), oils and fats, margarine, Purina Chow, molasses, honey, black pepper, cocoa, butter, and nonfat dried milk, most of them processed.

Although most of the cadmium we encounter comes from food, only small amounts are absorbed by the intestinal tract, and even those are lessened when the diet contains plenty of zinc. It is highly likely that the largest part of the cadmium in our bodies comes from air— from tobacco smoke, polluted air, and dusts—for most of the cadmium

Table VII-6. Cadmium in Human Livers and Kidneys

	Liver (ppm ash)	Kidney (ppm ash)	Ratio liver/kidney	Source of cadmium
Normal persons				
U.S.	180	2,940	.058	Food
Africa	50	810	.061	Food
Near East	140	1,600	.088	Food
Far East	480	4,200	.114	Food and air
Switzerland	190	2,300	.083	Food
Average	208	2,370	.081	
Hypertensive persons				
U.S.	190	4,220	.045	Food and water contaminated
Foreign	210	5,080	.041	
Average	200	4,650	.043	
Exposed workers				
Sweden	8,400	4,650	1.81	Dust in air inhaled
Great Britain	13,150	25,000	.526	Dust
Greece	8,800	5,500	1.600	Dust
England	33,000	6,200	5.323	Dust
Average	15,837	10,337	2.315	
Ouch-ouch disease	7,051	4,903	1.438	Food and water polluted

Note: This table shows where cadmium is deposited when it is taken into the body by mouth or by the lungs from the air. Airborne cadmium goes into the liver and the kidneys but mainly into the liver. Foodborne cadmium goes to the kidneys more than to the liver. People with poisoned livers are too sick to have hypertension. The rats in Table VII-3 that were fed 5 ppm cadmium in water had a ratio of .382.

SOURCES: Schroeder 1965, Friberg et al. 1971.

inhaled is absorbed directly from the lungs into the body. As little as 1 μg per day retained would build up a body burden of 14.6 mg in 40 years, over a third of the average amount, 38 mg, in all tissues. A pack of cigarettes contains 20 μg, of which 10 μg are absorbed from smoke. So it is easy to get enough cadmium to cause illness.

The pattern of storage of cadmium in the body differs according to the route of entry (Table VII-6). Workers exposed to heavy dusts, as in cadmium-nickel battery plants, had large amounts in their livers, with ratios of concentrations in liver to kidneys usually greater than 1.0. Total amounts stored, of course, were much larger, for the average liver weighs six times as much as the kidneys. Persons with high blood pressure had twice as much cadmium in their kidneys as did normal people, but no more in their livers. Japanese with ouch-ouch disease got their cadmium by mouth, and they actually had less than did those exposed to dusts in factories. The total amount of cadmium stored in the livers and kidneys of exposed workers was nearly 350 mg, the total in those organs of hypertensive persons was only 20 mg, and the total in normal persons was about 12 mg. Therefore, a little cadmium goes a long way.

Cadmium is a perfect example of an accumulative abnormal and subtly toxic trace metal in the environment causing widespread and serious human diseases, most of which are fatal. As such, cadmium is the worst of the bad actors among the metals.

VIII

ELEMENTS TOXIC AND NOT SO TOXIC

Three of the five toxic metals, cadmium, lead, and mercury, have been discussed at length. The other two, beryllium and antimony, do not make up an important public health problem at the present time, although they are very toxic substances in themselves and could become hazardous if industrial uses increased markedly. A nonmetal, arsenic, can be included among the toxic elements mainly because it causes cancer in human beings.

There are ten other industrial metals and one metalloid which have some degree of toxicity to mammals and to which the population is exposed. None of them constitute important health hazards at the present time, although they could if not watched. Present exposures are probably too low to result in disease, except in certain industrial and geological situations.

Potential exposures of the population to the toxic and slightly toxic metals are shown in Table VIII-1. There are many. Listed in the table are only those coming into direct contact with human skin, food, water, or air. In most cases, absorption through the skin is negligible, although some people are sensitive to nickel and zirconium, breaking out in a rash, and all people are probably sensitive to beryllium. These potential exposures are no cause for alarm, except in the cases of the first six metals on the list.

The sources for cadmium, lead, and mercury are repeated for pur-

Table VIII-1. Potential Human Exposures to Toxic Trace Metals

Metal or element	Source of exposure
	Toxic
Antimony	Pewter, britannia metal, rubber, flameproof textiles, dyes, paint, ceramics, type metal, medicines.
Arsenic	Insecticides, fungicides, weed killers, majolica, alloys, medicines, pigments.
Beryllium	Copper alloys, ceramics, rocket fuels.
Cadmium	Plating, alloys, pigments, insecticides, solders, cigarette smoke, plastics, rubber, galvanized iron.
Lead	Tetraethyl lead in gasoline (largest use), pipes, cisterns, paint, alloys, solders, glass, pottery glazes, rubber, plastics, insecticides, pewter.
Mercury	Paint, solders, fungistats, amalgams, drugs, disinfectants, jewelry. Organic mercury in seed dressings, paint.
	Slightly Toxic
Selenium	Coatings of stainless steel, glass, pigments, dyes, rubber, insecticides, plastics, dandruff preventives.
Tin	Tin plate, alloys, collapsible tubes, foil, plastics, rubber, fungicides, insecticides, drugs, toothpaste.
Barium	Alloys, paints, rubber, soap, linoleum, sugar, ceramics, insecticides, paper.
Germanium	Electroplating, alloys, catalyst for hydrogenation of coal.
Nickel	Alloys, steels, coins, enamels, ceramics, glass, catalyst for hydrogenation of oils.
Lanthanum	Lighter flints, weighting of silk and rayon.
Gallium	Alloys, glass-sealing compound.
Rhodium	Alloys, jewelry, plating.
Tungsten	Alloys, plating, flameproofing of cloth, pigments.
Bismuth	Pharmaceuticals, alloys.
Zirconium	Alloys, lighter flints, ceramics, pigment, plastics, catalyst, antiperspirants.

poses of perspective. Eight metals are used in paint, four in plastics, six in rubber, and six in ceramics, to which all of us are exposed. Fifteen are used in alloys and three in water pipes. Medicines and pharmaceuticals use eight of them, and three are applied directly to the skin. Four are incorporated in clothing. Six are in insecticides. We cannot avoid them.

Metal Cancers

Many people fear that metals cause cancer, and that contact with skin or with food in cans or pots and pans may lead to this dread disease in some mysterious way. The facts are clear, for much work has been done on this possibility.

Only one, arsenic, a metalloid, produces cancer in man when taken continuously into the stomach. Arsenic causes skin lesions and pigmentation. Some of the skin lesions appear on the palms and soles and become cancerous. There was an epidemic of arsenic cancer in Taiwan a few years ago, 52 cases appearing; deep wells with natural arsenic in water were responsible. In Cordoba, Argentina the water is arsenical and the cancer rate was high—23 percent of all deaths. Arsenic has not caused cancer in animals fed it for life.

Chromate dusts inhaled by workers have caused cancer of the lung. Inhaled dusts of nickel have caused cancer of the nasal passages and lungs in workers, and nickel carbonyl causes cancer of the lung. Nickel carbonyl is found in cigarette smoke, probably in auto exhausts burning nickel additives to gasoline, and probably in poorly combusted petroleum and coal. Beryllium dusts inhaled by workers have caused lung cancers.

No other metals absorbed by mouth or by air are known to cause cancer in human beings. Several given by mouth to experimental animals cause cancers: lead, selenium, rhodium, palladium. Although no one goes around injecting metals into people, a number of metals produce cancers when injected into animals: beryllium, cadmium, lead, nickel, chromium, cobalt, and titanium. Inhaled beryllium and nickel will produce cancers of the lung in animals, and inhaled lead, cancers of the kidneys. Therefore, nickel and arsenic are the ones to watch from this viewpoint.

Beryllium

This is probably the most toxic metal of all, for it causes a serious chronic lung disease which is incurable. Fewer than 900 cases have been reported, most of them in workers exposed to dusts or in members of their immediate families, although there are a few "neighborhood cases" in cities where beryllium is worked and the air is contaminated. There is little beryllium in food and water; the lung is the site of entry into the body and the place where the disease mainly occurs. Lung cancer occasionally occurs.

The air of nine U.S. cities and of four nonurban areas have small amounts of beryllium, mainly from high-beryl coals. Coal from different areas has from 0.1 ppm to 31 ppm beryllium; the estimated global emission is an annual 410 tons, but the amount varies widely according to the source of the coal. In industry, beryllium oxide, fluoride, and sulfate are toxic, whereas beryllium aluminum silicate, the semiprecious stone beryl, is inert. The air of Cape Canaveral, Florida, undoubtedly contained beryllium in toxic forms, for it was used in rocket fuels, and the exhaust contained 50 percent beryllium oxide, 40 percent beryllium fluoride, and 10 percent beryllium chloride. Chronic beryllium poisoning should eventually appear in that area in exposed persons if the winds were right and enough rockets were launched. It is no longer used for this purpose.

The human body contains very little beryllium on the average, about 30 micrograms, unless the person has been exposed to contaminated air. Beryllium is avidly accumulated by the human lung. When a particle gets into the skin through a scratch or a cut, it causes an ulcer which never heals until the particle is removed.

Beryllium is a bad actor, the most toxic of the environmental industrial metals. We should avoid it at all costs.

Antimony

As we have said, exposure to antimony is ancient; antimony sulfide (stibnite) has been used as a cosmetic by women for perhaps 6,000 years, in the form of kohl. Today potential exposures come from food and fluids in contact with ceramic enamels, pewter, and britannia metal; from clothing impregnated with antimony trioxide for flame-proofing; and from air in dusts from the wear of rubber and fumes of type metal. Contact with certain kinds of red rubber is also a source. Exposures, however, are low, except in industry.

Potassium antimony tartrate injected intravenously is the treatment of choice for liver flukes and other parasites. Large doses are commonly used; they produce heart toxicity and even sudden death. Liver damage has also been reported. In one Chinese series, mortality was 50 percent from treatment, which is no medical triumph.

Food contamination from acids dissolving enamel glazes has caused acute poisoning. Lemonade prepared in white enamel buckets can dissolve much antimony, enough to cause vomiting.

Miners, foundry workers, typesetters, rubber compounders, abrasive makers, and other exposed workers have a variety of toxic symptoms. The heart is often affected, with heart muscle degeneration. Transient pneumonia can come from inhalation. Skin disorders, eczema, dermatitis and irritation of the mucous membranes, pustular eruptions, and the like are common. We have seen mucous membrane irritation from a tube of red rubber in one nostril; a plastic tube in the other was nonirritant.

There is probably less than 6 mg antimony in the soft tissues and 2 mg in the bones of Reference Man.

Rats exposed for life to 5 ppm antimony in drinking water had shortened life-spans and longevities, some depression of fasting serum sugar and cholesterol, kidney disease, increasing hardening of the arteries, and a high incidence of heart attacks; there was little effect on growth. Life-spans of mice were not affected significantly.

Ambient urban air levels are not alarming, only four cities of 58 in the United States sporadically showing antimony, but because of its toxicity, antimony in air should be controlled. Intake from food is nearly 800 μg per day.

Arsenic

Arsenic, a metalloid, has enjoyed from antiquity a reputation characterized by actions both benevolent and malevolent. Among its virtues are its medicinal effects, said to be known to Hippocrates (460–357 b.c.), Aristotle (384–322 b.c.), his successor Theophrastus (370–288 b.c.), and Pliny the Elder (a.d. 23–79). As organic compounds first synthetized by Ehrlich in 1905, arsenicals have been effective in killing syphilis spirochetes, amoebae, sleeping sickness organisms, and many other parasites. About 1,500 compounds have been made. Among arsenic's mixed blessings are its destructive effects on insects, weeds, fungi, and pests.

Its principal vice is its toxicity, in certain forms, to man. It provides such a simple, inexpensive, and convenient vehicle in the art of homicide that the word "arsenic" has become synonymous with "poison." So bad is its reputation as a poison that a popular play is entitled *Arsenic and Old Lace*. However, the gentle but psychotic ladies portrayed in it, who repeatedly laid their visiting gentlemen to eternal rest, used strychnine!

The name was derived from the Arabic *Az-zernikh*, for orpiment (corruption of *auri pigmentum*, arsenic sulfide, also known as King's

Gold), and is attributed to Pedanios Dioscorides (ca. A.D. 50), a Greek physician who wrote *Materia Medica*. Deposits of arsenic do occur as the native element, but more frequently as compounds with many heavy metals, often as sulfides. When heated, arsenic is released from ores into the air and the flue soot. Cancer of the scrotum was an occupational disease of chimney sweeps; arsenic from coal was probably a factor. In 1250 Albertus Magnus is said to have obtained arsenic as the element, as did Paracelsus in 1520; in 1649 Schroeder (not this author) published two methods for its preparation.

Elemental arsenic is not toxic. Mountaineers in Styria and in the Alps of Switzerland and the Tyrol, for example, used to eat relatively large quantities of native arsenic, believing that it helped endurance at high altitudes, increased weight, strength, and appetite, and cleared the complexion.

I have seen mountaineers consume arsenic. In 1933 I was climbing in the Austrian Tyrol near the Zillerthal along with two medical students. The native guide took a handful of blackish gray crystalline material, said to be "arsen," from exposed rocks, ate a gram or two, and saved the rest for future consumption. A friend of mine, while a young man in Switzerland, observed mountaineers at home take a pinch of "arsen," a blackish gray crumbly material, spread it over bread and butter, and eat it as a tonic. What form it was in is not known, probably elemental arsenic. According to Professor R. Stigler of Going, Tyrol, "Arsenic was eaten by the mountain people in Styria and in Tyrol 40–50 years ago. They used to call arsenic 'Hiderich' or 'Hiderach' (Hüttenrauch). The arsenic eaters, especially in Styria and in the Zillerthal, were generally looking very well, cherry-cheeked and strong. But nowadays arsenic is no more used."

Arsenic is found everywhere on the earth's crust in fair-sized amounts, comparable to other toxic metals, at about 5 ppm. In seawater there are 2–5 ppb, which accounts for the fact that many seafoods have more arsenic than the FDA limits (2.6 ppm) allow: molluscs, crustacea—prawns and shrimp—fish, fish roe, cod liver oil, herring, even freshwater fish and most sea animals and plants. This arsenic is not toxic. The average man has 18 mg arsenic in his tissues, a little more than mercury and less than cadmium.

The second source of arsenic is a direct result of the Age of Metals. As long as 5,000 years ago, copper artifacts were made from copper sulf-arsenide ores; some of them contained as much as 12 percent arsenic. Smelting of copper and lead had been discovered by this time. Undoubtedly arsenic trioxide appeared in flue dusts and soot, for elemental arsenic oxidizes readily when heated, vaporizes slowly at 100°C, rapidly at 450°C, and sublimes at 613°C, below the melting points of copper (1083°), iron (1536°), silver (961°), lead (327°), and gold (1063°). Arsenic trioxide sublimes at 125°C, and many arsenical compounds are volatile at low temperatures. It is very likely that early workers in metals were exposed to arsenic trioxide when high smelting temperatures began to be used, for many artifacts of copper made later (2000 B.C.) were quite pure.

The use of arsenic has increased greatly during the past 90 years. In 1873 Great Britain produced 5,459 tons of arsenic trioxide as a by-product of the reduction of cobalt and nickel from ores. In 1902 it was used in the manufacture of glass for reducing the iron oxide in sand, and as pigments in aniline dyes, calico prints, and wallpapers, and small amounts were added to molten lead to produce spherical, hardened shot. An American ambassador to Italy was poisoned by arsenic-containing paint that fell from the ceiling into her food.

In 1926 production in the United States and Canada was 16,000 tons, and by other countries 11,000–12,000 tons; its use had been extended to wood preservatives, the insecticides lead and calcium arsenate, sheep-dips, flypaper, arsenical soaps, germicides, and rat poisons. By 1958 free-world production of arsenic trioxide was 40,000 tons, and by 1962, 55,500 tons. The entire output was a by-product of smelting arsenic-containing ores of copper, gold, lead, and other metals. Two companies accounted for all the arsenic produced in the United States. Because of the replacement of arsenical by organic insecticides, consumption of lead and calcium arsenates in this country has declined 57 percent in 10 years, but it has increased abroad.

The major sources of extraneous arsenic in vegetable foods are undoubtedly arsenical weed killers and pesticides. Lead and calcium arsenates and arsenites are firmly bound to the soil and are leached out slowly. Uncomtaminated garden soils are probably unusual. This source is decreasing with the substitution of organic pesticides. Trivalent arsenicals are also used in forestry to kill unhealthy or unproductive trees and to thin saplings, a process which will contaminate vegetation.

Accumulation of arsenic in human hair has recently aroused a minor international controversy. Smith, Forshufvud, and Wassen, from Glasgow and Göteborg, examined two samples attested to be of Napoleon's hair, finding in one 10.3 ppm and in the other 3.75 and

3.27 ppm. Analysis of sections indicated intermittent accumulations of arsenic in the growing hair. From this evidence it was deduced that Napoleon was deliberately poisoned by the British on St. Helena, a conclusion emphatically denied by Brock of England. Another British observer, Cawadias, in a detailed account of Napoleon's last illness, has proposed that he actually died as the result of a hepatic amoebic abscess rupturing into the stomach. A large dose of antimony was given less than two months before his death. Arsenicals are not illogical forms of treatment for intestinal amebiasis, although they are ineffective in hepatic abscesses. How the arsenic entered Napoleon's hair will doubtless remain unsettled; therapeutic administration of this commonly used "tonic" is not described in the various reports of his last illness.

The levels of arsenic in Napoleon's hair, however, were not especially abnormal for some modern environments. People living in two cities where lead, zinc, and copper smelting were taking place had more in their hair than did Napoleon (Table VIII-2). Hair also accumulated lead and cadmium, but not copper or zinc, for which there are excellent balancing mechanisms in the body, preventing excesses.

Everything is toxic in large enough amounts, and arsenic is no exception. Given in tea, it is an excellent, if painful, murder weapon, especially used by young relatives waiting for inheritances from rich old aunts. In smaller but still large doses it can produce cancer, but that takes a long time. Aside from these qualities, arsenic in the environment today appears to be harmless.

There has been an arsenic lobby of sorts conducted by a firm selling arsenical products for use in cattle and poultry feeds. Arsenic is a good killer of small organisms, and one of its greatest medicinal uses was to treat syphilis—with 606, or salvarsan, named supposedly because it took Ehrlich 606 tries to find a suitable compound. Arsenic added to feed makes chickens grow faster, probably by killing off some intestinal parasites. But in 1958 the Senate passed the Delaney Clause to the Food Additives Act, which prohibited the use of any food or feed additive which will cause cancer in animals or man. Arsenic is one such substance. Therefore a campaign of deliberate confusion was begun, and facts known since 1887 were vigorously, flatly, and repeatedly denied. That arsenic does not cause cancer in animals was stressed. The campaign went to such lengths as to deny that arsenic was responsible for the poisoning of 4,000 people in Manchester, England from contaminated beer; the arsenic came

Table VIII-2. Metals in Human Hair

Average levels of metals in human hair according to degree of industrial contamination.

City	Industry	Lead (ppm)	Cadmium (ppm)	Arsenic (ppm)	Copper (ppm)	Zinc (ppm)	No. persons examined
A	Lead and zinc mining and smelting	52.0	2.1	1.1	13.0	160.0	45
B	Lead and zinc smelting	20.0	1.6	4.0	12.0	145.2	25
C	Copper smelting	13.0	1.0	9.1	11.0	160.0	31
D	Government and commerce (near City B)	7.9	0.9	0.7	11.0	160.0	13
E	Education and farm trading	6.5	0.8	0.4	11.0	155.0	38
F	Small Vermont town, heavy traffic	18.4	2.3	—	16.1	169.6	129

Note: This table shows no substantial differences in copper and zinc because these essential trace metals have highly efficient homeostatic mechanisms in the body. It shows, however, the ranges of lead, cadmium, and arsenic in different industrial areas, and it indicates lead, zinc, and copper smelting as sources of cadmium and arsenic, and traffic as a source of lead and cadmium.
SOURCES: Hammer 1971, and personal observations.

from sugar, or from peat over which the malt was roasted, or from sulfuric acid, and the skin lesions were typical. Such outbreaks from contamination of grains, wine, and beer have been reported frequently. Present FDA regulations allow only very small arsenic residues in the meat of poultry and livestock; it is not fed for some time before slaughtering.

Arsenic is a poison and can cause cancer in man. Present exposures are probably not hazardous, although the soil of tobacco fields is heavily contaminated from the use of arsenates of lead or calcium as pesticides, and arsenic in tobacco has been blamed for lung cancer.

Selenium

In certain parts of the world selenium has been largely washed out of the soil by rainfall, and grasses contain little. In these areas, selenium deficiency occurs in sheep and in some cattle, for selenium is an essential trace element for mammals. Deficiencies have been found in newborn calves, lambs, foals, and rabbits, pigs, and birds in the U.S. in the Northeast, the Northwest, and parts of the South Atlantic Coast. Giving selenium prevents the disease, which affects normal growth, function of muscle and liver, and fertility.

Naturally the feed industry wants to add selenium to animal feeds for universal use, although in some areas high in selenium it might be toxic. Selenium, however, is included in the list of cancer-producing agents prohibited by the Delaney Clause, based on a study in 1943, one in 1946, one in 1960, two in 1961 and 1967, and one by the Trace Element Laboratory in 1970. This is a difficult decision for the FDA to make, for allowing selenium is against the law. Backed by the copper interests, who want to sell selenium, and the feed industry, which wants to add it to food at a profit, a deliberate campaign of repetitive denial was begun which amounted to vilification of good work and confusion in the minds of most officials and scientists not directly concerned. Not only is selenium not a cancer-causing agent in rats, according to these prevarications, selenium deficiency is actually a cause of human cancer! There is not a shred of evidence for this propaganda, which was designed to counteract the bad name of selenium. The copper interests are supporting projects with titles like "Selenium as a cancer preventative." Science thus becomes perverted for economic goals, and it is not the first time some unscrupulous scientists have been bought to protect an industry threatened by disclosures that its product is hazardous to health—the tobacco industry is a good example. It is dangerous to trust any industrial scientist talking about the product on which his job depends.

Selenite[1] fed to rats was highly toxic at 3 ppm selenium; selenate[2] was not. Selenite was only slightly toxic to mice fed it throughout their lives, but it caused tumors and cancers. Selenate produced large numbers of tumors, two-thirds of them malignant, in rats fed 3 ppm for life. Studies have indicated relatively high serum-cholesterol levels in rats given selenate and selenite, with a considerable increase in aortic fats and plaques. In mice, selenite was toxic to females but not to males, whereas selenate was not toxic at all. Because of the

---

1. A salt of selenious acid, H_2SeO_3.
2. A salt of selenic acid, H_2SeO_4.

extremely small requirements, one can calculate the toxicity as 100–300 times the required dose, or more.

Selenium is probably an essential trace element for man, and there is no evidence that it is carcinogenic in people exposed to large environmental concentrations. Excess selenium is toxic, however. Symptoms and signs reported include discolored and decayed teeth, yellow skin color, skin eruptions, chronic arthritis, atrophic brittle nails, edema, gastrointestinal disorders, and, in some cases, lassitude and partial or total loss of hair and nails. The older literature described severe congenital abnormalities in Colombia. In Oregon, Hadjimarkos has found a relation between selenium intake and dental caries in school children. Overt human toxicity has occurred only in persons living in seleniferous areas and consuming local food.

Cattle and sheep consuming seleniferous weeds with water-extractable selenium suffer weakness, lassitude, visual impairment, loss of appetite, paralysis with respiratory failure, and damaged viscera. Cattle, hogs, and horses consuming seleniferous grains and plants which collect and concentrate selenium suffer loss of hair and hooves, lassitude, anemia, and joint stiffness. The liver and the heart are severely injured. It is likely that selenium displaces sulfur in keratin, resulting in structural changes in hair, nails, and hooves. Its effectiveness in controlling dandruff probably lies in its extreme toxicity, poisoning the overgrowing cells on the surface of the scalp.

The first description of chronic selenium toxicity was written by Marco Polo, visiting the city of Su-Chan, China, in the province of Tangut, northeast of Lop-nor: "Over all the mountains of this province rhubarb is found in great abundance, and there merchants come to buy it, and carry it all over the world. Travellers, however, dare not visit those mountains with any cattle but those of the country, for a certain plant grows there which is so poisonous that cattle which eat it lose their hooves. The cattle of the country know it and eschew it. The people live by agriculture, and have not much trade. They are of a brown complexion. The whole of the province is healthy."

There are no reports of selenium deficiency in man, nor is there evidence that it occurs. Selenium was found in 210 samples of whole blood from donors in nineteen cities and towns in the United States. No blood was deficient. Food supplies in the United States are so widely distributed that it is doubtful that deficiencies or excesses of selenium in man occur in the general population.

Tin

Tin has a low order of toxicity. Rats fed 5 ppm in water for life had slightly decreased longevities, an increase in the incidence of fatty degeneration of the liver, some elevation of fasting serum sugar, and accumulation in tissues. Tin accumulates in the human heart with age.

Exposures are widespread through acid, canned foods which dissolve tin, although they are lower since lacquering of tin cans became standard practice. Some foreign cans are not lacquered and foods contain up to 200 ppm tin. Canning of foods began in the Napoleonic Wars as a result of the need to feed Napoleon's and the Duke of Marlborough's troops. The French discovered how to preserve food in glass; the British developed tin plate and the tin can. Tinned food still edible was discovered in an Arctic cache left in 1824 by Parry's expedition, which was looking for a Northwest Passage. The food contained 2,400 ppm tin, or 0.24%.

Alkyl (triethyl and tetraethyl) tin is highly toxic, but inorganic compounds of tin are not. It is probable that tin offers no hazard to health at present exposures, although the liver lesions in rats are disturbing.

Nickel

Inorganic nickel fed to rats and mice for life shortened life-span and longevity moderately, slightly affected reproduction, and produced some heart attacks. Nickel carbonyl, a combination of nickel and carbon monoxide, is carcinogenic in animals and man. Cigarette smoke contains this compound. It is formed at relatively low temperatures by passing the gas over finely divided nickel. Conditions for its formation probably exist in exhausts of motor vehicles burning gasoline containing nickel additives, and possibly in smokestacks where complete combustion of coal is not attained.

Nickel dusts are also carcinogenic in men exposed industrially. Because of this property, the possibility exists that nickel contributes to cancer of the lung. There is much nickel in the air of some cities, varying a hundredfold or more, with 0.03 to 0.12 $\mu g/m^3$ (micrograms per cubic meter), a pulmonary intake of 0.3 to 1.2 μg per day, or 110 to 430 μg per year.

Nickel carbonyl combines with lung tissue, probably for long periods of time. If the nickel in the air from burning of fossil fuels is found to be partly in the form of the carbonyl, it could be a potential hazard to health. Nickel emissions should be controlled.

Other Metals

Too little is known of the subtle toxicity of the other metals listed in Table VIII-1. When breathed or swallowed, barium shows little toxicity. It may accumulate in the lungs, giving rise to a mild lung disease. Added to diesel oil to control smoke in exhausts, it appears in high concentrations behind idling buses and trucks, and we probably inhale a lot of it when we are delayed in traffic behind an evil-smelling bus. Barium is also a natural contaminant of dusts, for there is plenty on the earth's surface. Huge amounts are swallowed for contrast in X-rays of the stomach and intestine.

Germanium is released from burning coal. Fed to mice and rats for life, it slightly shortened life-span. As discussed earlier, it inhibited the heading of rice, probably by displacing silicon; no human health effects have been described.

Tungsten interferes with the essential metal molybdenum, and causes changes in two enzyme systems in the liver of rats fed it in large amounts. Bismuth is quite insoluble, although toxic when injected. No adverse effects have followed large amounts drunk as "Soothing Pepto-Bismol." Zirconium is used as an antiperspirant, and some people get a rash from it. Rhodium caused cancers in mice. Little is known about lanthanum as a pollutant. Gallium is toxic to mice in terms of longevity. Boron is essential for higher plants.

Exposures are either too low for hazards, or toxicity is minimal as far as these eight metals are concerned, except in workers heavily exposed, who have a variety of symptoms (Table VIII-3).

Pollution of Air and Ocean

We don't have to worry too much about serious pollution of the air from these metals, and we don't have to worry at all about the oceans. Every year 410 tons of beryllium get into the air from coal; it all comes down on an earth with plenty around. Annually 9,600 tons enter the sea through weathering, of which 422 lbs remain in the water. Likewise 250 tons of antimony get into the air; about 0.15 percent of the amount weathered stays in seawater. About 700 tons of arsenic becomes airborne (Table VIII-4), 72,000 tons waterborne, and 108 tons seaborne. There are a total of 280 tons of tin in air from coal and 3,700

Table VIII-3. Diseases and Disorders in Workers Exposed to Metals

Metal	Type and extent of exposure	Effect on body
Aluminum	Heavy exposure to dusts	Irritation of nasal passages, eyes, and lungs
Antimony	Heavy exposure to dusts	Irritation of nasal passages, eyes, and lungs
	Ingestion	Toxic to heart. 100 mg fatal.
Arsenic	Ingestion	Skin lesions which can become cancerous; pigmentation. 120 mg fatal.
Beryllium	25 μg/m^3 in air	Beryllium disease (lungs). Cancers occasionally occur with beryllium disease.
	Particle entering skin through cut	Skin ulcers
	Ingestion	Toxic to skin and lungs. 1 mg fatal.
Cadmium	Ingestion	Toxic to kidney and lungs. 15 mg toxic, 35 mg very toxic.
Chromium	Long exposure to chromate dusts	Lung cancer; chronic skin ulcers.
Cobalt	Medium exposure to dusts	Irritation of nasal passages, eyes, and lungs
	In beer	Heart disease, increased red blood cells. 230 mg slightly toxic.
Lead	Ingestion	Toxic to brain and nerves. 2 mg/year toxic.
Magnesium	Heavy exposure to dusts	Irritation of nasal passages, eyes, and lungs; slow healing of wounds.
Manganese	Medium exposure to dusts	Pneumonia
	Heavy exposure to dusts	Parkinsonism; epilepsy.
Mercury	Ingestion	Toxic to brain and nerves. 2–22 g mercuric chloride fatal.
Nickel	Direct contact with skin	Eczema. Eyeglasses most common cause.
	Moderate exposure to nickel carbonyl	Lung cancer
	Heavy exposure to nickel dusts	Nose and lung cancers
Osmium	Heavy exposure to dusts	Irritation of nasal passages, eyes, and lungs. Toxic to eyes and nerves.
Platinum	Heavy exposure to dusts	Irritation of nasal passages, eyes, and lungs
	Direct contact with skin	Skin lesions

Table VIII-3 (cont.)

Metal	Type and extent of exposure	Effect on body
Selenium	Ingestion	10 mg/day toxic in time. Affects skin, nails, hair, lungs. Produces garlic odor in breath and perspiration.
Tellurium	Ingestion	Loss of perspiration; garlic breath, skin eruptions. 0.8 g fatal.
Thallium	Ingestion	Loss of hair. Toxic to nerves.
Tin	Heavy exposure to dusts	Irritation of nasal passages, eyes, and lungs
Tungsten	Heavy exposure to dusts	Irritation of nasal passages, eyes, and lungs
Vanadium	Heavy exposure to dusts	Irritation of nasal passages, eyes, and lungs
Zirconium	In antiperspirants	Chronic inflammation of skin in armpit

Note: There are no apparent effects from exposures to indium, titanium, palladium, molybdenum, gallium, aluminum, niobium, or tantalum. Silver is deposited in the skin and lungs without irritation. Copper salts produce skin irritation. Fatal doses are oral.
SOURCES: E. Browning 1969, Hamilton and Hardy 1949.

tons of nickel from coal and petroleum; of the 11,000 and 171,000 tons, respectively, entering the ocean in rivers and sediments, the sea retains 11 tons of tin and 8.55 tons of nickel.

These contaminants are negligible, except perhaps for beryllium, and are nothing like the amounts of lead emitted annually from car exhausts. Control of nickel emissions is most desirable, and control of beryllium and antimony highly necessary to protect local areas. There is no urgent need for control measures against the others.

Cost of Pollution by Industry and Weathering

At this point, it might be intellectually profitable to look at the economic losses resulting from man's need to burn fossil fuels. For seven common metals present in coal and oil, the global air emissions amount to more than one billion dollars a year (Table VIII-5). The largest loss by far is in aluminum, the next in lead. The least loss is in cadmium.

Economic losses of these same metals by weathering are over 100 billion dollars annually, the largest being in aluminum, copper, zinc, and nickel. These metals are deposited in alluvial sediments and on

Table VIII-4. Principal Sources of Human Exposures to Toxic Trace Metals

Metal	Source	Conc. in source (ppm)	Amount in smoke
Barium	Diesel fuel additive	200–750	12,000 μg/m³
Beryllium	Rocket exhausts	[1]	[1]
	Coal	<0.1–31.0	410 tons/year for coal containing 1.5–2.5 ppm
Boron	Coal	25–116	
	Iron baghouse dust	500–540	
Cadmium	Copper smelting emissions		6.7 kg/ton Cu produced 2,091 metric tons in air/year, world.
	Zinc refinery dusts	4,200	
Selenium	Coal	0.1–7.38	0.001 μg/m³ air, Cambridge, Mass.
Vanadium	Coal	176	
	Oil	100–600	
	Oil fly ash	630,000[2]	
	Oil fly ash	47,000[3]	
	Residue oil	23–500	
Arsenic	Producers-gas	300,000	
	Dust near cotton gins	300	6.4 g/bale cotton produced
	Coal	0.08–16	327–6,440 tons in air/year, world
Nickel	Asbestos	1,500–1,800	
	Coal ash	3–10,000	
	Crude oil	55	
	Asphaltene	245	
	Diesel exhausts		
	Fuel	2	
	Exhaust coke	10	
	Particulates	10,000[4]	
	"	1,000[5]	
	"	500[6]	

Table VIII-4 (cont.)

Metal	Source	Conc. in source (ppm)	Amount in smoke
Nickel (cont'd)			
	Coal	130–690	3,700 tons in air/year, world
Chromates	Coal		18–500 $\mu g/m^3$
	Asbestos	1,500	

1. Classified information.
2. Fuel from Venezuela or Iran.
3. Fuel from Texas or California.
4. No load at 1,400 rpm.
5. No load at 1,800 rpm.
6. Half load at 1,800 rpm.
SOURCES: Schroeder 1969, 1970–71.

Table VIII-5. Economic Losses from Industry and Weathering

Cost of metals emitted into air annually from burning fossil fuels and added to seas by weathering, 1972 New York prices.

Metal	Price/lb ($)	Burned		Weathering	
		Added to air (thousands of metric tons)	Cost (thousands of dollars)	Added to seas (thousands of metric tons)	Cost (thousands of dollars)
Copper	0.525	2.1	2,425	330	381,150
Zinc	0.175	7	2,695	800	308,000
Nickel	0.80	3.7	6,512	171	300,960
Lead	0.155	400	136,400	131	44,671
Mercury	2.17	10.6	50,604	3.5	16,709
Cadmium	2.00	0.042*	184	300	1,320,000
Tin	1.81	0.28	1,115	11	43,802
Aluminum	0.29	1,400	893,200	154,000	98,252,000
Total			1,093,135		100,667,292

* Plus 11,400,000 lb from copper and zinc smelting and refining and from incinerators, or $22,800,000.
SOURCES: *New York Times*, Financial Sections, January 1973, and Tables IV-1 and 2 above.

the sea floor near the land. With small dead single-celled organisms as nuclei, manganese and iron slowly form nodules with onionlike structures, scavenging from seawater nickel, cobalt, and copper in enormous quantities. In the Pacific Ocean alone there are an estimated 1.5 trillion tons of metals in nodules, forming at the rate of 10 billion tons a year, or two to eight times the annual world consumptions of fifteen common industrial metals. It is possible to reclaim these metals, for mining on a pilot scale has begun. If only 10 percent could be mined economically, at the present rate of consumption of metals there would be enough manganese, cobalt, nickel, gallium, vanadium, and other metals to last 150,000 to 400,000 years.

IX

PURE FOOD IS POOR FOOD

Five of the twelve leading causes of death in the United States are directly or partly nutritional in origin— heart diseases from hardening of the arteries, hypertension, strokes from the same two causes, diabetes, and cirrhosis of the liver. Trace metals are involved in all five and probably involved in another, emphysema. In two others, congenital defects and cancer, trace elements may be involved. Of the other four, accidents, suicides, and birth injuries are preventable, and pneumonia is usually curable. Therefore, nearly 94 percent of the fatal diseases in this country involve trace elements in one way or another, too much or too little, and 70.7 percent are directly involved with them.

Not only deaths, but chronic diseases are affected by metals. The activity of nearly eight million people is restricted because of diseases due to metals. Lead affects all urban dwellers subclinically and children are especially susceptible; probably seven to ten million children are affected. Cadmium probably affects 23 to 40 million adults. At least 250,000 babies born every year in the United States have congenital defects (6%), and more careful studies place the number higher (14.1%); some of these cases are probably caused by metals in the mother.

One important consideration is that mammals, probably including man, that are in a state of marginal or poor nutrition with respect to vitamins and trace elements are much more susceptible to infectious diseases from viruses and bacteria and to the toxic effects of chemicals, including toxic metals, than are those in an optimal state of nutrition. For this reason, to neglect the question of nutritional adequacy in any discussion of toxicity would be to present only a part of the picture.

We can illustrate this point by citing the results of lifetime exposures of several hundred rats and mice to lead. Lead caused deaths in young rats and shortened life-spans in both rats and mice when their diet was deficient in chromium. When the experiments were repeated with the same diet but with traces of chromium added, the toxicity to young rats was abolished and the prolonged effects considerably lessened. If the animals had been in a state of optimal nutrition, the toxicity of lead would have been hard to detect.

Contamination of our food and beverages with toxic trace metals is sporadic and irregular, and in most cases is caused by industrial practices. On the whole, however, our food is fairly free of these contaminants. Contact with containers can cause considerable solution of metals by acid foods on long standing. All modern canned food contains tin, but there is much less than formerly since in the U. S. cans must now be lacquered and the tops crimped to prevent the contents from coming in contact with the solder. Many imported cans are not lacquered, and there is considerable tin in canned anchovies, sardines, and other fish from foreign sources, and in edible oils dispensed in tin cans. Tin is not very toxic, however. Canned asparagus takes up tin and the consumer so likes the taste that when asparagus is preserved in glass, tin is added to make it resemble the original (contaminated) food—but not the fresh. One can detect tin in canned food by adding the juice of black currants or blackberries to the contents; a blue or violet color denotes tin or iron. The juice of Persian berries turns yellow, forming a tin pigment used in confectionery.

Acid beverages and foods will dissolve cadmium from galvanized iron containers and cadmium-plated vessels and produce acute illness. They will also dissolve lead from poorly glazed pottery and cause poisoning. They have dissolved antimony from enamelware and produced toxicity. But in general, the food we buy is not contaminated, and only rarely contains the toxic metals which act in malnutrition. We get the metals from air and water, rather than the supermarket.

What is the nutritional situation for the American public today? According to the overpositive statements of one expert, there is nothing to worry about. (His laboratory is heavily supported by the food industry, I am told, making him a mouthpiece for the industry and thus not credible. No dog bites the hand that feeds him.) All other signs, however, point to a cause for concern.

A compilation of government surveys and private studies made by Ralph Nader's group is quite explicit in its conclusions. Every ten years the U.S. Department of Agriculture conducts a nationwide household food survey, based on seven nutrients: protein, the metals calcium and iron, and four vitamins. The nation's diet deteriorated between 1955 and 1965, only 50% of the population having "good" diets in the latter year, compared to 60% in the former. The worst deficiencies were in protein, vitamin A, and vitamin C. The adequacy of diets declined with income, but even in affluent families 9 percent had poor diets and only 63% had good diets. Although only seven nutrients were estimated, it is clear from other studies that a diet low in several nutrients may be low in many.

In 1969, an article compiling the results of 32 separate studies showed that about half the population (48 percent) consumed less than adequate amounts of four micronutrients (iron, thiamin, riboflavin, and niacin); and 15 percent (or 32 million Americans) consumed less than two thirds of adequate amounts. Using actual measurements of vitamins in the urine, 13 to 30 percent were deficient, according to 29 studies. That means that 25 million Americans did not eat half the recommended daily allowances of one or more nutrients, and 50 million were deficient in at least one. The largest of such studies, involving 88,000 persons in ten states, showed similar findings; 44 percent were deficient in one or more of six nutrients, and 13 percent in two or more. Again the effect of poverty was obvious, 60 percent of lower-income and 28 percent of higher-income groups being deficient. All these surveys point to one inescapable conclusion: the Great American Diet is not what it should be.

The trouble with such surveys is that deficiencies of zinc, chromium, magnesium, manganese, copper, molybdenum, and other essential elements and of several neglected vitamins are not measured or listed, although presumably they are present. A low intake of protein means a low intake of zinc, for example. The reason for those omissions is clear. The Food and Nutrition Board of the National Research Council is composed, like most such boards, of cautious, conservative people who refuse to stick their necks out for fear they may be wrong. No decision is better than a questionable decision, for no one can criticize no decision. No individual, of course, stands up to be counted, because, as so well expressed by the psychiatrist, Frank Egloff, M.D., "Feelings of confidence drive out feelings of confidence. Feelings of competence induce feelings of incompetence." Thus it has been with the Food and Nutrition Board, which took 35 years to recognize vitamin E as essential for man after being shown to be essential for mammals, and 30 years to recognize vitamin B_6 for the same purpose, 15 years after human deficiency had been described. Such dilatory tactics suggest that the board cannot make up its bureaucratic mind as to man's mammality.

The well-known incompetence of the Food and Nutrition Board would be merely a source of tolerant amusement if the board were given the attention it deserves. But it is not. As an advisory body it is the ultimate authority for food and vitamin manufacturers, nutritionists, and physicians, and seems to set the pace for, but not always agree with, countless other boards.

Obviously, no two boards can agree on recommended daily allowances of nutrients. The American board has changed its standards for four vitamins by 12.5 to 31 percent each time it met, and reduced its level of calories by 12.5 percent over a period of 15 years. Boards of other countries vary in their recommendations for vitamins and calories by 50 to 71 percent, a symptom of the state of the art. It is doubtful that man is that variable.

Not only is the American board inconsistent, but it does untold harm by its omissions. When it omits a recommendation for a nutrient because it won't make up its mind, that nutrient is neglected by vitamin manufacturers, bakers, and food processors. What isn't mentioned is assumed to be not needed by man, even when it is a vital requirement of all living things. Thus, large numbers of people may suffer from some degree of deficiency.

So it was with the trace metals. The board has set standards for one, iron, which is added to refined flour. (Even so, iron deficiency has been found in 24 to 38 percent of the population at older ages. The iron formerly added was in a form only 20 percent of which was absorbed by the intestine; it was a dollar a pound cheaper than the absorbable form, now used.) Zinc was first discovered in plants in 1854, and was shown to be essential for the growth of a lower plant in 1869 and for all higher plants in 1926. In 1934, laboratory animals were first made zinc-deficient, and in 1940 an enzyme present in all animals which was concerned with carbon dioxide was found to contain zinc. During the next 27 years zinc was shown to be involved in iron metabolism, to promote wound healing even in healthy young men, to be necessary for reproduction, for growth, for the formation of the genetic substance of cells (DNA and RNA), for the formation of the eye, for prevention of a fatal skin disease of pigs, to be a constituent of many enzymes, and to prevent symptoms of poor blood supply in the legs from hardening of the arteries. Zinc deficiency was described in human beings, and many people were found deficient in zinc—pregnant women, women taking the pill, Iranian and Egyptian dwarfs, and patients suffering from infections, alcoholism, liver damage, ulcers of the leg from poor circulation or sickle cell anemia, heart attacks, cystic fibrosis, loss of sense of smell, Mongolism, and idiocy. Most of these cases improved when zinc was given. Although most micronutrients are probably interdependent, if one had to choose the most important one it would be zinc.

The American board finally recognized that zinc was an essential trace metal in 1968, but set no level of requirement. In January 1973 the FDA approved zinc at 15 mg per day in vitamin-mineral supplements as "optional." So zinc is still largely neglected, and it is most difficult to obtain tablets of zinc for self-medication, except in health food stores. One such preparation is labeled "Organic Zinc Supplement," the ZINC in large letters. It costs $2.00 for 100 tablets. The small print says that four a day supply 2 mg zinc, a number of B-vitamins, and, strangely, 1 mg of sulfur. The requirement for zinc is at least 15 mg per day. One preparation contains 26 mg zinc and considerable magnesium, a dosage which is acceptable. The large drug houses will not market zinc; there is no profit in it. Physicians now must persuade their pharmacists to make up capsules of zinc salts, which are costly—one drugstore charges 12½ cents each.

Zinc represents a major breakthrough in nutrition and in the treatment of some chronic diseases. It is not toxic. As much as 150 mg a day by mouth has been given to patients for periods of several years. It does not accumulate in tissues. It benefits only those who are deficient, but zinc deficiency of slight degree is widespread in this country. Animals raised for profit and for laboratory experiments are fed large doses for optimal growth and performance: for example (parts per million of Purina Laboratory Chows) rats 58–100, guinea pigs 122, mice 24, monkeys 20, rabbits 33, dogs 41–178. On this basis, man would get 8–15 ppm.

Neglect of all other trace metals was also the policy of the high moguls of nutrition on the Food and Nutrition Board until 1968, when copper, chromium, and manganese, but not molybdenum, were recognized reluctantly. No recommendations were made in the face of a great many animal and some human experiments. Copper was first identified in plants in 1817 and it is essential for all forms of life; its necessity for plants has been known since 1917 and for animals since 1928. Several enzymes are known to require copper for activity. Malnourished infants may be deficient and become anemic; otherwise dietary deficiency seldom, if ever, occurs in humans. Plenty of copper seems to come from copper water pipes and cooking ware. The French' and Indians, who use copper pans and pots exclusively, tin their copper to prevent absorption by food, apparently believing that copper is bad for one. Diseases suspected to be caused by too much copper are arthritis and scleroderma; high serum levels occur in women who are pregnant or taking the pill, and are associated with some cancers and some psychoses. Monoamine oxidase, a copper enzyme governing mood, is inhibited by drugs which chelate, or bind, copper, including the mood elevators and those designed to remove copper from the body. The copper transport system is governed by a gene for copper; when hereditarily missing, copper is absorbed in large amounts, causing liver and brain disease and early death. Normal animals get 11–21 ppm copper in prepared food and man gets 3–5 ppm.

Manganese was found in plants and animals in 1923, and its essentiality proven by 1936. It is a constituent of a few enzymes, and is concerned with fat and cholesterol and with joints. Animals are given 40 to 115 ppm in food for their health—no chances are taken—whereas man gets 2–4 ppm. No one knows why. It is obviously not toxic by mouth. We believe that many Americans are somewhat deficient but we cannot prove it.

Chromium was found in plants in 1911 and in all living things by 1948. One enzyme concerned with sugar and phosphorus has chromium in its molecule. It was found to be essential for sugar metabolism (1959) and for cholesterol metabolism (1968), and deficiency is associated with atherosclerosis, the cause of heart attacks.

Molybdenum was discovered in all animal tissue in 1932, and in 1953 was shown to be a constituent of an enzyme governing uric acid. Deficiency may be linked to tooth decay, cancer of the esophagus, and kidney stones, and excesses may lead to copper deficiency (in cattle and sheep at least). One can grow striped wool in black sheep by raising or lowering the intake of molybdenum; the copper is necessary for pigmentation.

Cobalt was regularly found in plants in 1841, but not until 1948, after twenty years of research, was it found to be a constituent of vitamin B_{12}, the anti-anemia factor. B_{12} requirements for man are the smallest of any known active substance, vitamin, mineral, or hormone —one microgram a day, or a millionth of a gram, containing 4.34 percent cobalt. Vitamin B_{12} is made by bacteria in the sea and deep lakes, and in the intestinal tracts of mammals; the only source is food of animal origin, such as milk, eggs, meat, cheese, and liver, a fact which should, but does not, cause concern to strict vegetarians until they become anemic. Deficiency not only induces pernicious anemia but also degeneration of the spinal cord. There is plenty of cobalt in the body other than in B_{12}, and it may possibly have other functions. Large surpluses put in beer to hold the foam have caused severe heart disease, and smaller amounts ingested by children have led to thyroid disease and overproduction of red blood cells. Foods for animals contain 0.13 to 0.53 ppm cobalt, monkeys needing the most.

We can bring all these dietary needs together for comparison by analyzing two days' worth of an ample hospital diet (and also depending on analyses of others). Table IX-1 shows the results. Let us assume that the whole day's intake was adequate for a healthy person (and that assumption is questionable), and that ideally each meal should supply one third of the day's concentration of a metal. If we then call deficient any meal that does not supply 50 percent of that amount, we can see that breakfast was deficient in vanadium and chromium, lunch was deficient in molybdenum, and dinner was deficient in manganese. Cobalt, copper, zinc, selenium, strontium, and magnesium were fairly evenly distributed among the three meals, but the total intakes for zinc and magnesium were less than those recommended for a good diet.

There seem to be three groups of metals with respect to amounts in the diet. Magnesium is a bulk metal. Zinc and iron (15 mg a day) are the next in abundance; the zinc value found here is low. Occurring in 1–3.2 mg amounts are vanadium, manganese, copper, and strontium; in less than 0.5 mg amounts are chromium, cobalt, nickel, molybdenum, and selenium. These dietary intakes are roughly proportional to body stores (chapter 2).

The types of foods supplying these trace metals and elements are given in Table IX-2, based on a hundred or more analyses of each metal. By no means are the metals evenly distributed in all foods. Seafood is deficient in manganese and molybdenum, but six oysters can have as much as 400 mg zinc and usually have over 100 mg. (The be-

Table IX-1. Amounts of Essential Trace Elements in 2-Day Hospital Diets

Element	Diet no.	Breakfast (μg)	Lunch (μg)	Dinner (μg)	Total (μg/day)	Concentration (μg/g)
Vanadium	1	*15*	220	930	1,165	0.46
Chromium	3	*34*	75	275	384	0.18
Manganese	2	866	1,239	*297*	2,402	1.42
Cobalt	2	53	50	62	165	0.09
Nickel	1	245	137	90	472	0.19
Copper	1	903	1,214	1,061	3,178	1.26
Zinc	2	2,269	3,737	2,488	8,494[1]	5.0
Molybdenum	3	65	*46*	259	370	0.17
Selenium	3	31	12	20	63	0.029
Strontium	3	350	1,030	690	2,070	0.97
Magnesium	3	64,200	90,300	43,600	198,100[2]	93

Diet 1 was an institutional diet of fruit juice, oatmeal, griddle cakes, and toast for breakfast; roast pork, potatoes, sauerkraut, and applesauce for lunch; and macaroni and cheese, fruit salad, bread, and jello for dinner; with tea, coffee, and milk. It weighed 2,526 grams and contained 2,040 calories.

Diet 2 was a hospital diet of fruit juices, cereal, eggs, bacon, toast and coffee for breakfast; beef or salmon, potatoes, peas, tomatoes, spinach, pie, and bread for lunch; and soups, turkey, chicken, fruit, gelatin, and salad for dinner; with plenty of milk. It weighed 1,697 grams and contained 2,230 calories.

Diet 3 was similar to *Diet 2*, with lunches of lamb or beef, cottage cheese, potatoes, fruit, hot chocolate, chocolate pudding, and bread; and dinners of roast lamb and pork, potatoes, peas, beans, chocolate cake, pie, and breads. It weighed 2,140 grams and contained 2,150 calories.

Note: Amounts in italics are those where concentrations are less than one sixth of the total day's intake. They indicate deficient meals, which must be made up for by meals having excesses.
1. Recommended Daily Allowance 15 mg. This diet 43.4% deficient.
2. RDA 400 mg. This diet 50.5% deficient.

lief that oysters are aphrodisiacs may have some basis of fact; a man can lose 1–2 mg zinc in a single seminal emission.) Meats are deficient in vanadium, manganese, and nickel; dairy products are deficient in vanadium and nickel. Assorted vegetables contain less than half the estimated requirements of selenium, and fruits are generally poor in vanadium, chromium, zinc, and selenium. Oils and fats are low in zinc, molybdenum, strontium, and magnesium. Nuts and cereals are ade-

Table IX-2. Types of Food Containing Essential Micronutrients
Average values (μg/g).

Element	Sea-food	Meat	Cereals & grains	Dairy products	Vege-tables	Fruits	Nuts	Oils & fats
Vanadium	1.7	*0.1*	1.1	*0.01*	1.5	*0.1*	0.7	3–40
Chromium	0.11	0.14	0.2	0.1	0.5	*0.02*	0.2	0.1
Manganese	*0.05*	*0.2*	7.0	0.7	2.5	1.0	11.1	1.8
Cobalt	1.6	0.2	0.4	0.1	0.1	0.1	0.3	0.4
Nickel	0.6	*0.02*	0.8	*0.03*	0.5	0.15	—	1.1
Copper	1.5	3.9	2.0	1.8	1.2	0.8	14.8	4.6
Zinc	17.5	30.6	17.7	4.2	*1.6*–10.7	*0.5*	34.2	*2.4*
Molybdenum	*0.01*	2.1	2.5	0.2	0.1–4.8	0.1	—	*0*
Selenium	1.0	0.92	0.15	0.06	*0.0*	*0.0*	0.03	1.0
Strontium	9.6	1.2	5.0	0.5	0.6–6.7	0.7	60	*0.3*
Magnesium	348	267	805	157	220	78	1,970	*7*
Vitamin B$_6$	2.0	3.0	4.0	0.7	2.1	0.6	5.0	—

Note: Concentrations in italics are those which represent less than 50% of the mean in the whole diet (see last column of Table IX-1).

quate or contain surplus amounts of all trace elements; coming from seeds, they contain the elements needed for the growth of the seed until the plant forms roots by which it can obtain them from soil.

With a little imagination one can see what happens when a food is partitioned into its fatty and nonfatty fractions, as we do so frequently with milk. The fat is deficient in magnesium, molybdenum, and zinc, but keeps its vanadium, chromium, manganese, cobalt, nickel, copper, and selenium. Magnesium, however, is needed to metabolize the fat and it must come from somewhere else, such as nuts and cereals. So butter, peanut butter, and bread would make a good combination for digesting the butter.

If we ate the raw and cooked foods analyzed, we would get adequate amounts of vitamins and minerals, enough to metabolize the food. Fat, a natural food, contains the micronutrients necessary for its metabolism by the body. But today, in order to preserve food, we refine it, purify it, process it, scald it, freeze it, and fraction it with little regard to what happens to the micronutrients. These processes have provided or could provide almost everyone with adequate calories but not necessarily with adequate micronutrients to digest and use them properly. Thus, deficiencies of several vitamins and minerals are world-wide, and every older nutritional survey has detected them in many countries.

There is nothing new about this form of malnutrition. The invention and spread of machines to polish rice during the last century was followed by an epidemic of beri-beri throughout the Orient; losses of bran removed vitamin B_1, zinc, and many other minerals and vitamins. Long voyages at sea without citrous fruits and fresh vegetables caused scurvy due to vitamin C deficiency. Before cod liver oil, rickets was prevalent in all nothern cities; insufficient sunlight and insufficient vitamin D were responsible. Night blindness from vitamin A deficiency was common in healthy soldiers in World War II. Some degree of anemia from iron deficiency is prevalent among menstruating women and teen-age girls. Zinc deficiency of some degree is being discovered everywhere, now that we are measuring for it.

Undoubtedly some vitamin and mineral deficiencies are unseen, not apparent, not visible, manifest only in chronic poor health or in athero-sclerosis, or perhaps allergies, obesity, arthritis, and other common ailments of civilized man. We do not know, but we can suspect that that is so.

The essential trace elements are much more important than the vitamins. They cannot be synthesized, as can the vitamins (by bacteria and plants), but must be present in the environment and taken into foods therefrom. They cannot be metabolized, although they often change their valences. Their only sources are the earth's crust and sea-water. Without them life would cease to exist. It is probable that every balanced food carries in it the micronutrients necessary for its metabolism, both elemental and organic; less than these amounts can result in abnormal metabolism of the food.

When a food is highly refined and the micronutrients are largely removed, the calories are called "empty," lacking in nutrients for their metabolism. Sugar—white sugar—is an example. It has essentially no vitamins and mere traces of minerals, and any of these substances necessary for the metabolism of the sugar must be provided by other unpurified foods or by pills. This fact is inadvertently but beautifully illustrated by a misleading series of advertisements from Sugar In-formation, sponsored by the International Sugar Research Institute, if that is its present name (it changes). One shows a smiling child and a peanut butter and jelly sandwich. The blurb states that sugar is a food and implies that it is a whole food, for the vitamin and iron content of the peanut butter sandwich is listed. Not the sugar, for that is zero. Another shows a girl—also smiling—behind a huge dish of ice cream covered with caramel sauce—empty—and topped by a strawberry. "You need vitamins, minerals, proteins, fats and carbohydrates" the blurb declaims (true enough). "And it just so happens that sugar is the best tasting carbohydrate. *Sugar. It isn't just good flavor; it's good food,*" say the clever Madison Avenue writers. It just so happens that sugar is made of empty calories, and has no vitamins, minerals, proteins, or fats in it. "You should take in a balanced diet," the ad declaims so rightly. "And that's where sugar comes in." We know that that's where sugar comes in, to unbalance the diet by 20 percent.

Refined sugar makes up about 20 percent of the diet of the average American. Therefore, his diet is unbalanced to start with.

Other carbohydrates when refined, especially the grains and cereals, also unbalance his diet—or your diet. In Table IX-3 are shown the

Table IX-3. Nutrients Lost in Refining Whole Products

Percentages lost in refinement of whole raw wheat, raw sugar, unpolished rice, whole corn, and whole milk.

Nutrient	Wheat flour	Sugar, refined	Rice, polished	Corn starch	Milk, fat-free
Ash	76	88	—	—	32
Calcium	60	—	—	—	—
Phosphorus	71	—	—	—	—
Magnesium	85	98	83	97	6
Chromium	40	93	75	72	>50
Manganese	86	89	45	93	100
Iron	76	—	—	—	—
Cobalt	89	95	38	37	0
Copper	68	83	26	31	0
Zinc	78	98	75	91	14
Molybdenum	48	100	—	—	90
Selenium	16	100	—	100	88
Strontium	95	96	—	—	0
Vitamin B$_6$	72	100	69	87	—

SOURCE: Schroeder 1971.

actual measurements of the percentage losses from the refining of
wheat into flour (losses 40 to 80 percent of the amount of trace metals
in whole wheat), polished rice (losses 26 to 83 percent of the trace
elements in unpolished rice), and corn starch (losses 31 to 100 percent
of trace elements in whole corn). Vitamin losses which have been
measured show 86 percent for vitamin E, 72 to 81 percent for the B
vitamins, and 50 percent for pantothenic acid and 67 percent for
folacin in the refining of wheat into flour. The residue, known as mill-
feeds, goes to livestock and poultry as a rich vitamin and mineral
supplement.

When we fractionate milk, which is a perfect food, we make it im-
perfect. The skimmed—or fat-free—milk contains most of the magnesium, cobalt, copper, and zinc of the original whole milk, but has
lost manganese, molybdenum, selenium, and half the chromium. It has
also lost vitamins A, D, and E. Thus, fat-free milk is no longer a whole
food, but is unbalanced; to balance it means that we must add the
original butter to it. Or what is better and cheaper, throw it out and
buy whole milk.

Therefore, a diet with plenty of white bread, refined cereals, white
rice, and corn starch, perhaps 40 percent of our calorie intake and the
stuff of sandwiches, pies, rolls, and pastry, supplies only 30 to 40 percent of the necessary micronutrients. This diet is unbalanced, and the
body must get what it needs from other foods. Wheat flour is not *all*
"empty calories"; only 60 to 80 percent of it is, and must be com-
pensated for from other sources.

Alcoholic drinks such as whiskey, vodka, gin, and rum contain no
vitamins, but are not entirely devoid of metals (Table IX-4). They undoubtedly come from the distilling plant or the aging kegs, for metals
are not volatile and not distillable. Whiskey has a little of everything.
Wines, of course, have what was in the grapes and beer what was in
the yeast, malt, hops, and water. So these calories are not strictly
empty. But since there is too little zinc to metabolize the alcohol in
hard liquor, it can be considered empty for all practical purposes.
Adding snacks rich in zinc, such as nuts, liver paté, shrimp, or smoked
oysters, to the cocktail hour will help utilize the micronutrients in the
liquor which would otherwise be lost. We receive 10 to 20 percent of
our total calories from alcohol, of which 7.1 percent of an average daily
intake of 2,150 calories is from beer. We consume 125.14 lb of sugar
a year, or 155 grams, 620 calories a day, or 28.8 percent of a 2,150-
calorie diet (19.4 percent of a diet of 3,200 calories).

Table IX-4. Metal Concentrations of Alcoholic Beverages (mg/liter)

	Whiskey	Wine, domestic	Wine, Italian	Beer
Vanadium	—	—	—	0.004
Manganese	—	0.5–15	—	0.1
Iron	0.05	10	—	0.4
Cobalt	<0.025	0.03–0.08	0.5–12	0.03
Copper	0.3–0.45	0.25	0.38–0.88	0.38–0.6
Zinc	0.05–0.15	0.04–5.0	0.5–5.6	0.28–0.38
Molybdenum	0.17	0.05	—	0.01–0.06
Magnesium	1.0–4.5	—	135	—
Nickel	—	—	—	0.01–0.06
Lead	0	0.35	0.08–0.66	0.1
Selenium	—	0.05	—	0.19

Note: The amount of lead in wine is more than the average daily intake by mouth. Some wines will provide plenty of zinc.
SOURCES: *Ten-State Nutrition Survey;* National Distillers; and personal observations.

The situation looks very bad when examined from this viewpoint. The American Diet lacks at least 30 percent of its micronutrients when it includes unrefined grains, and is 40 to 50 percent empty—"naked," as some say—when white flour, cornmeal, and rice are used. For the diet to be balanced, the micronutrients must come from other foods or from a bottle.

We can see how hard it is to cover these deficiencies by diet alone. In Table IX-5 are calculated the concentrations needed in the rest of the diet when 10 to 50 percent of the diet is composed of empty calories. Taking the concentrations in Table IX-2 for different classes of food, the table lists those which are higher than the amount required by a diet of 40 percent empty calories.

To be specific, your diet of 2,500 calories contains the average amount of refined sugar (620 calories) and alcohol (630 calories), and you like white grains, which average 70 percent empty and of which you eat 800 calories. You will have to fill half your intake, 1,250 calories, plus 560 calories (70 percent of 800), a total of 1,810 calories, with 24 micronutrients from other sources. Where do you get them?

Table IX-5. Compensating for Empty Calories

Average concentrations of nutrients necessary in the remainder of the diet to achieve recommended levels when part of the diet consists of empty calories.

Nutrient	μg of nutrient needed per gram of diet Percent empty calories						Foods containing nutrient needed to make up for empty calories
	0%	10%	20%	30%	40%	50%	
Calcium	800	889	1,000	1,144	1,333	1,600	Milk, cheese, milk products
Magnesium	350	389	438	500	583	700	Grains, nuts
Vanadium	0.56	0.56	0.63	0.71	0.83	1.0	Grains, vegetables, oils
Chromium	0.15	0.17	0.19	0.21	0.25	0.3	Vegetables
Manganese	1.0	1.11	1.25	1.43	1.67	2.0	Grains, vegetables, nuts, fats
Iron	15	16.7	18.8	21.4	25	30	Meats, grains
Cobalt	0.1	0.11	0.13	0.14	0.17	0.2	Seafood, meats, grains, nuts, fats
Copper	1.25	1.39	1.56	1.79	2.08	2.5	Nuts, oils and fats, meats
Zinc	15	16.7	18.8	21.4	25	30	Meats, nuts
Molybdenum	0.2	0.22	0.25	0.29	0.33	0.4	Meats, grains, legumes
Selenium	0.04	0.04	0.05	0.06	0.07	0.08	Seafood, meats, grains, fats
Nickel	0.2	0.22	0.25	0.29	0.33	0.4	Grains, vegetables, fats
Vitamin B_6	2	2.2	2.5	2.9	3.3	4.0	Grains, nuts

Note: Weight of diet is 1,000 g solid food. Column 5 shows that if 40% of the calories were empty, one would have to find an extra 583 mg magnesium, 0.25 mg chromium, 1.67 mg manganese, 25 mg zinc, etc., from the foods listed in the last column. The figure 40% empty calories is a conservative one.

The Food and Nutrition Board has recognized for many years that 24 nutrients are removed from flour during its refining, and recommends that four be put back, calling it "enriched." So you get adequate B_1, B_2, B_3, and iron in your white flour. But the rest of the 24?

A look at Table IX-2 will show that you must eat seeds: seed grains and nuts, and they must be unrefined, whole, untreated except for grinding. Wheat, rice, barley, rye, oats, millet, and nuts supply you with excess vanadium, manganese, copper, zinc, and magnesium. If you need more zinc you can get it from animal protein. Oils will give you vanadium; seafoods and meats selenium, cobalt, and molybdenum. Only vegetables, on the whole, will provide chromium. Fresh fruits have little of any trace element, although their seeds are rich.

Nickel is shown in the tables because it may be essential for man. The nickel in fat may come from its use as a catalyst for hardening unsaturated fats into saturated fats. Strontium is shown for the same reason.

Four more of the micronutrients refined from wheat are bulk elements—sodium, potassium, phosphorus, and calcium—and we get plenty of them from other foods. Six other vitamins are removed to the extent of 23 to 86 percent, and not put back, but we don't know too much about their deficiencies in the population; four are crucial and two may not matter.

Therefore, we can see that the Food and Nutrition Board, which was designed to protect our health, is not wholly incompetent to do so. Incompetence is only 20 out of 24 things, or 83 percent. A batting average of .167 hardly rates the major leagues.

We can do something about this situation. There are only three courses open. (*a*) We can eat unprocessed and unrefined foods—if we can get them. Try any supermarket. It is hard. (*b*) We can make the food refiners and processors put back what they take out. They will do it too. They want to sell. But that won't work for the unstable vitamins: pantothenic acid, folacin, C, E, B_6. (*c*) We can take pills. But in this case we must get all the vitamins and minerals in adequate amounts. An adequate amount, to be safe, is the dietary content or the recommended dose.

Dark brown sugar contains adequate amounts of the trace elements necessary for metabolism of the sugar, for it has molasses added. Raw sugar as we get it is already quite refined and is generally inadequate. Molasses contains the nutrients taken out of sugar, and is rich in chromium, manganese, copper, zinc, and molybdenum, and contains some selenium.

Our fats such as lard, shortening, and oils are also refined, and the chromium necessary for their proper utilization is largely lost. Our purified lard has very little vitamin E in it. Butter has plenty of chromium, manganese, cobalt, copper, and molybdenum.

All of us are probably deficient in one or more essential micronutrients. To maintain a balanced diet, we need analyses on each of 24 micronutrients in all our foods and a computer, and no one has those facilities at his daily beck and call.

Among Americans deficiency of chromium in human tissues begins in childhood and progresses slowly through life. Atherosclerosis, or hardening of the arteries, is the result of chromium deficiency. There are other factors operating in this common disease, but it has two manifestations: (*a*) elevated cholesterol and other fatty substances in the blood, and (*b*) an inability to handle glucose efficiently—a mild state of diabetes. The cholesterol and other fats are deposited in the walls of arteries, narrowing them and sometimes closing them off. The result in the arteries of the heart is a heart attack, which kills middle-aged men and older people; in the brain a stroke, a rupture of the artery, a brain hemorrhage; in the legs, gangrene of the feet. We call this "ischemic vascular disease," "ischemic" meaning too little flow of blood. It kills more than half of us Americans before our time.

Analyses were made of the aorta (the large central artery) and seven other tissues of some 200 foreign and 200 American patients. There was virtually no chromium in the arteries of people dying of heart attacks, but some chromium in those of people dying of accidents. Chromium was present in the tissue of almost every foreigner, but was missing in 17 to 60 percent of tissues of Americans at all ages. There was 4.5 times as much chromium in the bodies of Orientals, 5.6 times as much in Near Easterners, and 1.6 times as much in the bodies of Africans as in those of Americans. Heart attacks are not a major cause of death in most foreign areas. In Bangkok, chromium was high and there was very little atherosclerosis and a low incidence of diabetes.

These analyses point to a deficiency of blood-vessel chromium in heart attacks. The deficiency, as we have shown, is dietary.

In about half the cases, feeding chromium salts results in the lowering of blood cholesterol and the normalizing of sugar tolerance. Chromium in inorganic compounds, however, is poorly absorbed, and it is known that the form found in Nature is in an organic complex known as the glucose tolerance factor, or GTF. Serious efforts are being made to purify the substance and make it; when that is done the day of specific treatment of atherosclerosis will dawn. The situation resembles cobalt and vitamin B_{12}. Synthesis of B_{12} introduced the control of pernicious anemia by a vitamin. The human body can make B_{12} in the colon, but little is absorbed. The human body probably can make GTF in small amounts, but not enough to really count. And not at all when there is little or no chromium.

Chromium deficiency in rats duplicates the human picture, with elevated blood cholesterol, elevated blood sugar, and deposits of fatty substance in the aorta. It is prevented by feeding chromium, which lowers blood sugar and blood cholesterol.

There may be another influence in the development of atherosclerosis, vitamin B$_6$ deficiency. Deficient monkeys have weakened areas in their blood vessels, the precursors of deposits. When fed cholesterol, the monkeys show fatty deposits in these areas. B$_6$ deficiency, thus, may be a silent disease of the population, going hand in glove with chromium deficiency and preparing the way to serious vascular disease.

The vested interests, which are enormous (meat packaging runs four billion dollars a year), seem to be confused about the Great American Diet. There is one ex-scientist at a prestigious university who has sold himself to the food trade, which supports his laboratory at a high cost. He repeatedly plays the same old record to the news media —that the American Diet is the best in the world—but no one in the know believes a word he says, only the unknowing public. Another excellent scientist at the same institution tells the truth, but he has trouble finding a forum. The sugar industry spent large sums of money on proving that cyclamates caused cancer, while keeping under the rug the findings that several sugars also cause it; cyclamates are now banned but the sugars are not. The result: more sugar is sold, and the more sugar sold, the more unbalanced becomes the American Diet. And the more people get hardening of the arteries, for there is just as good evidence that white sugar causes heart attacks as that saturated fats do. (As we have shown, the common factor in sugar, flour, and fat is lack of chromium.)

Not only the food industry is confused and defensive; so is bureaucracy. As this chapter was being written, the latest joker arrived. The Food and Drug Administration solved the problem of the deficient Great American Diet by requiring labeling of all packaged foods, the labels to include the contents and allowances of the nineteen vitamins and minerals now officially recognized as essential by the Food and Nutrition Board. Our diet is actually more deficient than it was formerly thought to be because the RDAs for calcium, phosphorus, and magnesium have been raised (U.S., Dept. of Health, Education, and Welfare, "Food Labeling," p. 2124). Furthermore, zinc (15 mg a day) and copper (2 mg a day) at last have been officially recognized as essential, and biotin and pantothenic acid have been added. The FDA now declares that we need twelve vitamins and seven minerals— chromium, manganese, cobalt, molybdenum, and selenium remain in bureaucratic limbo. The reader will note that only three of the essential trace metals are considered worthy of inclusion (see Table IX-6).

Table IX-6. Recommended Daily Allowances (RDA) of Vitamins and Minerals, 1968, 1973.

Adult man, mg

Substance	1968	1973
Bulk Minerals		
Calcium	800	1,000
Magnesium	350	400
Phosphorus	800	1,000
Silicon	N.R.	N.R.
Trace Minerals		
Vanadium	N.R.	N.R.
Chromium	R, no RDA	N.R.
Manganese	R, no RDA	R, no RDA
Iron	10	18
Cobalt	N.R.	N.R.
Copper	R, no RDA	2
Zinc	R, no RDA	15
Molybdenum	N.R.	N.R.
Selenium	R, no RDA	N.R.
Iodine	0.11	0.15
Fluorine	1	no RDA
Vitamins		
A, International Units	5,000	5,000
C	60	60
D, International Units	100	optional (400)
E, International Units	30	30
B_1	1.2	1.5
B_2	1.7	1.7
B_3	14	20
B_6	2	2
B_{12} (μg)	6	6
Folic Acid	0.1	0.4
Biotin	no RDA	0.3
Pantothenic Acid	no RDA	10

N.R. Not recognized.
R, no RDA. Recognized, no recommendation made.
SOURCES: National Academy of Sciences 1968; U.S., Dept. of Health, Education, and Welfare, "Food Labeling."

These new regulations will be beneficial in that the intelligent consumer can buy food knowing the protein, fat, carbohydrate, caloric, cholesterol, salt, vitamin, and mineral content. They will encourage the food industry to add vitamins and minerals to its products to fulfill the RDAS in order to advertise and sell the food. But they will do harm in omitting chromium and manganese from the list, for our foods will contain only what remains after processing.

Especially unfortunate is the regulations' virtual prohibition of using manganese, chromium, and molybdenum in dietary supplements. Manganese is used now in supplements, and is mentioned in the regulations (para. 12, p. 2146: "The essential mineral nutrients are: calcium, chlorine, iron, magnesium, phosphorus, potassium, sodium, sulfur, copper, fluorine, iodine, manganese, and zinc") but not included in the tables of requirements. Why not? If any chronic disease is the result of manganese deficiency, we can expect a marked increase in incidence after these regulations go into effect. The incidence of chromium deficiency and atherosclerosis in the U.S. is now nearly 100 percent, and cannot go much higher.

All of this just goes to show how little people know about the subject.

A retired nutritionist with long government experience was talking to me about the present situation. Finally she said, "I don't think we will ever get Government to act. It may be that the only solution is for people to patronize Health Food Stores so heavily that industry will come around. Health Food faddists and cranks will learn more and more, and the public may become educated. Then they will demand action by government." Perhaps.

We can do several things useful to our health. We can avoid white flour and its products, white sugar and its products ("Pure Cane" or "Pure Beet"), and white shortening. We can insist that what is taken out of food in the preserving is put back. The new regulations on labeling micronutrient content of foods go into effect December 31, 1974; we can insist that they include, along with the 19 officially accepted micronutrients, the equally important or more important trace metals chromium, manganese, cobalt, molybdenum, and selenium. In that way, we can know what we are eating.

X

CONTROL OF METAL POLLUTION

When we view the whole problem of metal pollution in its many ramifications, it appears so huge as to seem hopeless of solution. We live in an age of metals and we cannot change that. Nor will it ever be changed drastically unless the human race goes backward to wood, iron, and stone in a hippielike existence.

When we view each problem making up the whole individually, we see glimmers of light. The only way to control pollution by metals is to attack each metal by itself, and choke off environmental emissions at the source. If we can. If we cannot, there are other methods devisable.

Let us look at the "water factor," for example. The water factor is something in soft or acid waters which promotes death from heart attacks and high blood pressure. It sounds strange, but it is true in Japan, the U.S., Canada, Great Britain, Sweden, Holland, and South America. After extensive surveys and much computer time, it has finally been decided that nothing measurable in the water itself is responsible for heart disease, but that the corrosiveness of the water on the pipes does the damage. Your morning cup of coffee contains the metals dissolved from pipes overnight, unless you flush them out. The metal most likely in the pipes is cadmium, which is attacked by soft water. Cadmium will induce high blood pressure, which leads to heart attacks. The British are solving this problem directly by hardening their municipal waters. We should do the same by simple treatment with lime and magnesium salts, which makes corrosive water more alkaline and harder. If that were done for all cities and towns with acid waters and less than 50 ppm hardness, it is likely that the death rate from heart attacks would decline and equalize, over the country. To illustrate the effect of the water factor on heart death rates, the variation in mortality figures for 94 cities was analyzed. For males the highest rate was more than twice the lowest; for females it was three times the lowest. We could expect a reduction in death rates of about 25 percent for this one cause, if this kind of "pollution" were controlled.

Cadmium emissions can be controlled as suggested in chapter 7. Smokestack emissions from zinc, lead, and copper smelters should be prevented by methods already in use. The fact that cadmium does not occur in detectable amounts in the air of many cities means that we do not have to have it in any city's air. Electroplaters responsible for discharging cadmium washings should be made to clean their drip water before pouring it down the drain. Cadmium-nickel battery makers should not discharge cadmium wastes into waters or onto dumps. The price of cadmium would partly repay efforts to recycle and reclaim the metal. Smoke from incinerators must be cleaned of cadmium. To prevent solution of cadmium in streams, tailings of zinc mines must be guarded, a very difficult task. Cadmium alloys or coatings should not come in contact with food.

It is simple today to measure cadmium in food and fluids. If found, the source can be tracked down by anyone with imagination. Analyses are rapid and sensitivity is increasing yearly. Soon we will be able to put a small bit of food, tissue, or fluid into a tube, heat it instantaneously, and read off the concentration of 30 trace elements at very little cost.

Among the ancillary effects of abatement of cadmium in air is abatement of smoke particulates, lead, mercury, and other trace elements in coal and oil. Among the ancillary effects of abatement of cadmium from water is better conversion of sewage to elemental products which can be used as fertilizer, thus avoiding the use of high-cadmium superphosphates. When we go after one toxic contaminant, a lot of others are removed as well.

Mercury can be recovered from plant discharges by known methods. The amount dumped into waters has decreased markedly in two years, and mercury in rivers and lakes should no longer be a problem, provided laws are enforced. Mercury can be removed from coal-burning stacks by abatement of smoke. Not until this world converts to atomic power will the source of mercury, carbon dioxide, smoke, and many metals disappear entirely. Enforcement of laws against the use of methyl mercury as seed dressings and against the dumping of mercury in any form should solve the present problem and remove the threat to health.

Antimony must be removed from smoke. People must be taught not to let acid foods and fluids stand in enamelware.

Beryllium must be removed from smoke.

Solution of the problem of lead toxicity is easy. Remove the lead from gasoline and 90% of the source will vanish. Remove the cadmium from smelter exhausts and the lead will go with it. Replace lead water pipes with plastic or copper and find a substitute as a plasticizer and stabilizer for plastic pipes, and most of the lead will be gone. Disallow lead paint, as Australia does, and another source is removed at no hardship.

Nickel also should be abated from smokestacks; it will be removed with the cadmium, lead, and mercury and with the particulates. Nickel compounds should not be added to gasoline for fear of the formation of nickel carbonyl in the exhaust system.

When the air is cleaned up, many contaminants will disappear. As to how clean the air should be, there is a wide difference of opinion among the experts. The Environmental Protection Administration farmed out the job of deciding air quality standards and limits for metallic pollutants. It was quite clear that the Mitre Corporation, which got the contract, knew less than nothing about the subject, nor, admittedly, did the charming, bright young woman who interviewed me. That is the way the government works, and major decisions depend on the person interviewed.

At any rate she listed my guesses along with those from a federal scientist and a state-employed scientist. The two government workers gave much higher safe levels than I did; apparently they did not take into account whether or not a metal was absorbed from the lung into the body or accumulated in the lung, nor how toxic the metal was in itself. They did not realize that 5 $\mu g/m^3$ (5 micrograms per cubic meter) cadmium in air would result in an accumulation in the body of 50 μg per day—about as much as the food intake—or 18 mg in a year, half the body burden. One of them gave a bad guess of 0.2 $\mu g/m^3$ for beryllium; at this level the lungs would collect 730 μg in a year and produce beryllosis in 10 years. Another guess of 10 $\mu g/m^3$ for lead would cause accumulation of 36.5 mg per year and would invite lead toxicity in less than 10 years. Levels for the essential metals were proposed; as we know by now, these metals do no harm, and need be limited only for the sake of cleanliness.

If these recommendations are accepted as standards by the government, we are in for a bad time. We will have to rename the Clean Air Act the Dirty Air Act, for such standards would allow 100 times as much beryllium, 37 times as much cadmium, twice as much lead, 25 times as much mercury, and 24 times as much nickel as we find at present.

The metals found to accumulate in human lungs with age are the toxic one, beryllium, the slightly toxic ones, barium and tin, and the nontoxic ones, chromium, aluminum, iron, strontium, titanium, and vanadium. Beryllium is the one to watch. The metals which are absorbed into the body via the lungs are barium, beryllium, cadmium, nickel, lead, tin, and strontium, and the essential ones, chromium and manganese. Cadmium is another one to watch. We cannot insist too strongly that it and beryllium be controlled at levels as near zero as possible.

There is a great temptation for government to use as limiting values for metals in air those already established to protect workers exposed to metal dusts and fumes. Industrial exposures of eight hours' duration for 40 years have little or nothing to do with constant exposures from the cradle to the grave for the whole population. Threshold limiting values were the best guesses at the time, and were designed to prevent acute or chronic disease in workers. No attention is paid to subtle, or recondite, toxicity, for that has seldom been recognized in industry.

Even when the limits in workers' air are enforced, certain industries carry excessively high mortality rates from cancer. Mechanics and repairmen, foremen, metal and other craftsmen, engine and construction machinery operators, carpenters and cabinet makers, transport and public utility workers have significantly more lung cancers than do less-skilled workers or professional and white collar workers; so do painters and plasterers, metal operatives, drivers and delivery men, workers in manufacturing industries (but not mine operatives and laborers), and laborers in metal manufacturing. Some of these workers also have more stomach cancers and more cancers of all types. In all these occupations, exposures to metals or metal dusts and fumes are

obvious. Service workers and construction and outdoor laborers are
free from this hazard, having no more cancers than other groups.
Therefore the threshold limits do not necessarily protect against cancer, an example of recondite toxicity, and are useless, misleading, and
potentially harmful for the general population. Table X-1 shows what
will happen if some of these limits, designed for the dirty situations
which prevail, are adopted.

Air-quality standards were proposed for eighteen metals. On the
assumption that half the metal breathed will be absorbed into the
body, the standards proposed for sixteen of those metals would lead
in ten years to accumulations exceeding the average body burdens,
and if the air were allowed to get that polluted, we would expect
endemic poisoning with beryllium, cadmium, and lead to a greater
extent than now, perhaps with mercury and nickel, and possibly with
arsenic. These limits are unsafe, and show the abysmal ignorance of
the men who proposed them. No, we must do better than that, at least
as well as we have done to date.

Vermont has some of the strictest environmental control laws in
the nation. One law forbids discharging anything extraneous into
streams and rivers without a temporary state permit. Consequently,
the waters are becoming purer each year, and the Connecticut River
is now fairly clean and full of game fish. With several fish ladders,
we can have Atlantic salmon again.

Federal drinking water standards are now quite acceptable, except for cadmium, and few large municipalities exceed them, although
a number of small towns do. The situation is changing for the better.
Potentially, however, some of these standards seem to allow a greater
proportion of toxic metals to be assimilated from water than advisable
when the total intake from all sources is considered.

In Table X-2 are shown the allowable limits of six elements in
water, and the proportions of their intakes from waters compared to
all sources, which could occur if water were just barely acceptable—
just under the wire. The allowable limits of cadmium undoubtedly
should be lowered to 5 ppb, no antimony or beryllium allowed, and
the lead level, which is tolerable, enforced. But these limits are little
cause for concern and are seldom reached at present.

One must be reasonable about any standard for air or water, and
not throttle industry. But we can be stricter than we are now, inasmuch
as 80 to 90 percent of the nation's waters have much less than the
allowable limits and 50 percent of the cities have no detectable toxic

Table X-1. Limits on Concentrations of Airborne Metals

Some proposed "safe" limits and the consequences if they are adopted.

Metal	Working limit[1] (μg/m³)	Accumulation[2] in body/10 yrs (mg)	Proposed limits[3] (μg/m³) A B	Accumulation[4] in body/10 yrs (mg) A B	"Safe"[5] limits (μg/m³)	Accumulation[6] in body/10 yrs (mg)	Reference Man body burden[7] (mg)
Beryllium	0.02	0.024	0.2 0.001	*7.3* 0.036	0	0	0.03[8,9]
Cadmium	1–2	12–24	1 5	*36* *182*	0	0	38[9]
Lead	2	24	10 5	*365* *182*	0.5	18	120[9]
Antimony	—	—	—	—	0	0	8[9]
Mercury	0.5	6	0.1 2	3.65 73	0.1	3.65	13[9]
Nickel	10	*120*	10 10	*365* *365*	0.03	0.11	10
Tin	20	*255*	20 20	*730* *730*	0.03	0.11	17[8]
Vanadium	0.5–5	*6–60*	10 5	*365* *182*	10	*365*	22[8]
Chromium	5–10	*60–120*	10 10	*365* *365*	1	*36*	1.8[8]
Arsenic	5	*60*	10 10	*365* *365*	0.15	0.5	18

1. Presently adopted industrial working limits (Threshold Limiting Values, TLV), exposure 8 hours/day, 40 hours/week, 30 years.
2. Calculated accumulation at working limit in 10 years.
3. Limits proposed by consultants A and B, for ambient air quality (AAQ), 24 hours/day, 7 days/wk, 365 days/year, ages 0–70+ yrs.
4. Calculated accumulations at these proposed levels in 10 years.
5. "Safe" limits as proposed by this author, based on knowledge of toxicities.
6. Calculated accumulations in body at these safe levels.
7. Body burden of American Reference Man at present, for comparison with calculated accumulations.
8. Accumulates in lung tissue with age at present AAQ exposures.
9. Accumulates in body with age at present exposures from air, water, and food.

Note: Figures in italics approximate or exceed average body burdens. As such, exposures are unrealistic, and if allowed, would cause toxicity during a lifetime exposure to beryllium, cadmium, lead, and nickel. Elements such as arsenic and mercury are fairly rapidly excreted, but could build up to toxic levels at the higher exposures.

Accumulations are calculated on the generous assumption that half the amount of an element inspired in air remains in the body, mostly by absorption in the lungs. Thus, an air level of 10 μg/m³ in 20 m³ breathed per day would mean 100 μg retained in lung, or 365,000 μg (365 mg) in 10 years.

SOURCES: Working Paper, Mitre Corporation; and personal calculations.

Table X-2. Limits on Concentrations of Waterborne Metals

Maximal allowable concentrations of trace elements in drinking water, compared to total intakes from food and air.

Elements in water	Allowable limit in water (μg/liter)	Intake from food and air (μg/day)	Proportion of total intake from water (%)	Remarks
Cadmium	10	75	21.1	Water is major source
Lead	50	450	18.2	Water is major source after air. 5–10% absorbed.
Antimony	Trace	<1,000	?	—
Beryllium	Trace	0.01	?	—
Mercury	5	20	33.3	Water is major source
Chromium as chromates	50	200	33.3	Water is major source
Nickel	20	400	9.1	
Arsenic	50	900	10	

Note: All of these metals but mercury and arsenic are poorly absorbed from the intestinal tract. We assume that 2 liters of water are drunk per person per day, and that this water is barely acceptable for drinking purposes. In other words, these balances represent the maximal allowable amounts from water. It is probable, however, that most Americans do not drink 2 liters of tap water a day. The average consumption of bottled soft drinks is 568 ml per day, or nearly 19 ounces, and much milk is drunk.

Sources: U.S., Dept. of Health, Education and Welfare, *Public Health Service Drinking Water Standards, 1962;* and personal experiments and observations.

element in their air. When a waterborne toxic metal is found to come from natural geological sources we must leave it alone, but when it comes from industrial sources we must clean it up. Just as man's technology is not equal to cleaning the stratosphere of dust from a volcanic explosion, so is it inadequate to clean out a cadmium-zinc deposit or natural arsenic in water-bearing rocks.

Some people will say that we do not yet know enough about the exact mechanisms of disease from pollutants, and until we do, we have no justification for controlling them. That, of course, is nonsense. A poison is a poison, and most of us do not need to know exactly how it poisons before we begin to avoid it. There is enough knowledge today about the metallic pollutants of our civilization to give us every reason to act now.

The longer we delay and argue and promote indecision the more of the world's children, and our children, will die or be permanently damaged. And the more of us oldsters will suffer and die before our time from diseases due to our own carelessness.

BIBLIOGRAPHY

Bertine, K. K., and Goldberg, E. D. *Science* 173 (1971): 233.

Bowen, H. J. M. *Trace Elements in Biochemistry.* New York: Academic Press, 1966.

Browning, E. *Toxicity of Industrial Metals,* 2d ed. New York: Appleton-Century-Crofts, 1969.

Cowdry, E. V., and Steinberg, F. U., eds. *The Care of the Geriatric Patient,* 4th ed. St. Louis: C. V. Mosby, 1971.

Environmental Protection Agency. "Position on the Health Effects of Airborne Lead." November 1972 (typewritten).

Friberg, L., Piscator, M., and Nordberg, G. *Cadmium in the Environment.* Cleveland: Chemical Rubber Company, 1971.

Hamilton, A., and Hardy, H. L. *Industrial Toxicology,* 2d ed. New York: Paul B. Hoeber, 1949.

Hammer, D. I. "Trace Metals in Hair." *Journal of the American Medical Association* 215 (1971): 384.

Hernberg, S., and Nordman, H. "Study on the Hazards to Health of Persistent Substances in Water." Draft document on lead. Helsinki: World Health Organization, EURO 3109 W, 1972 (typewritten).

Howell, G. P. "Physiological Data for Reference Man." Chapter III of *Report of the Task Group on Reference Man,* W. S. Snyder, chm. Oxford and New York: Pergamon Press, in preparation.

Litton Reports. *Air Pollution Aspects of Beryllium, Boron, Selenium, Arsenic.* Bethesda, Md.: Litton Systems Inc., Environmental Systems Division, 1969.

National Academy of Sciences. *Recommended Dietary Allowances,* 7th ed. Publication 1694. Washington, D.C., 1968.

New York Times. Financial sections, January 1973.

Schroeder, H. A. "Cadmium as a Factor in Hypertension." *Journal of Chronic Diseases* 18 (1965): 647–56.

――――. *Air Quality Criteria*. Monographs #7012–17 on Barium, Cadmium, Vanadium, Nickel, Chromium. Washington, D.C.: American Petroleum Institute, 1970–71.

――――. "Losses of Vitamins and Trace Minerals Resulting from Processing and Preservation of Foods." *American Journal of Clinical Nutrition* 24 (1971): 562–73.

――――, and Kraemer, L. "Cardiovascular Mortality, Municipal Water and Corrosion." Submitted for publication.

Stitt, F. W., Clayton, D. G., Crawford, M. D., and Morris, J. N. "Clinical and Biochemical Indicators of Cardiovascular Disease among Men Living in Hard and Soft Water Areas." *Lancet,* Jan. 20, 1973: 122–26.

Tipton, I. H. "Gross and Elemental Content of Reference Man." Chapter II of *Report of the Task Group on Reference Man,* W. S. Snyder, chm. Oxford and New York: Pergamon Press, in preparation.

Underwood, E. J. *Trace Elements in Human and Animal Nutrition,* 3d ed. New York: Academic Press, 1971.

U. S. Dept. of Health, Education and Welfare. "Food Labeling." Food and Drug Administration. *Federal Register* 38 (Jan. 19, 1973): 2124–64.

――――. *Public Health Service Drinking Water Standards, 1962.* Publication 956. Washington, D.C.: Government Printing Office, 1962.

――――. *Ten-State Nutrition Survey, 1968–70.* Health Services and Mental Health Administration. Publication No. (HSM) 72–8130–34. Atlanta: Center for Disease Control, 1972.

U. S., Dept. of Interior. *Mineral Facts and Problems, 1967.* Bureau of Mines. Washington, D.C.: 1970.

――――. *Mercury in the Environment.* Geological Survey Professional Paper 713. Washington, D.C.: Government Printing Office, 1970.

――――. *Reconnaissance of Selected Minor Elements in Surface Waters of the United States, October 1970.* Geological Survey Circular 643, in cooperation with the Bureau of Sport Fisheries and Wildlife. Washington, D.C., 1971.

Wallace, R. A., Fulkerson, W., Shults, W. D., and Lyon, W. S. *Mercury in the Environment—The Human Element.* ORNL NSF-EP-1 ORNL-NSF, Environmental Program. Oak Ridge, Tenn.: Oak Ridge National Laboratory, 1971.

INDEX

Agriculture Department's food survey, 113

air pollution: by burning fossil fuels, 28–31, 107, 109–10; and cadmium, 89; control of, 132–34; by industry, 28–29; from lead, 36–58; by miscellaneous metals, 105, 107

air quality standards, 132–33, 135

alcoholic beverages, 122–23

alloying metals, 18

American Academy of Pediatrics, 43

American Medical Association, 39

animal evolution, 9–13

antimony: in air and seawater, 105; ancient use of, 17; control of use of, 132; exposure to, 96–97

arsenic: ancient use, 17; as cancer cause, 95; pollution by, 105; toxicity of, 97–101

arteries, hardening of, 125–27

atherosclerosis, 125–27

automobile exhaust, pollution from, 37, 55–58, 132

Bailey, James L., 37

barium, 105

BBC (British Broadcasting Corp.), 43

BEAP (Biologic Effects of Atmospheric Pollutants), 44, 46

Begich, Nick, 64

beryllium: cancer caused by, 95–96; control of pollution by, 132–33; pollution by, 105

bismuth, 105

blood pressure, cadmium's effect on, 81–88

boron, 105

Boyle, Robert H., 76

bronze, 16–17

B₁₂ vitamin, 117

cadmium, 73–92: and blood pressure, 81–88; control of, 89–91, 131; in fish, 75–79; human exposure to, 89–92; in kidneys and liver, 91–92

calories, empty, 120–24

cancer: caused by metals, 95; lung, caused by beryllium, 95–96; of workers exposed to metals, 133–34

canned foods, 112

Chaucer, 17–18

chromium, 116, 125–26

children, lead poisoning of, 43–44, 53–56

Chow, Tsaihwa J., 38, 45

cigarettes' cadmium content, 81

cities, airborne lead in, 40–42, 46–47

civilization's use of metals, 15–25

Clean Air Act, 133

cobalt, 19, 117

columbium, 20

containers, food, 112

contamination: environmental, 26–35; of food and beverages, 112; types of, 4; see also pollution

control of pollution, 130–37

copper: in diet, 116; history of man's use of, 15–16

cost of pollution, 107, 109–10

Cotzias, George, 20

daily allowances: of nutrients, 114; of vitamins and minerals, 128
Delaney Clause, 100, 102
Department of Agriculture, 113
Department of Justice, 75–77
diet, micronutrients in, 111–29
dirt in streets, lead in, 55
diseases: from excess of elements in body, 21–24; heart, 87–88, 130–31; lung, 95–96; nutritional origin of, 111–29; of workers exposed to metals, 106–7, 133–34
"dragon's teeth." *See* cadmium
drinking water: cadmium pollution of, 88; fluoridation of, 23; and heart disease, 88, 130–31; standards for trace metals in, 134–36
duPont Corp., 40, 45, 47

earth's crust, elements in, 21–24
economic losses from pollution, 107, 109–10
Egloff, Frank, 114
Ehrlich, Paul, 97, 100
E. I. duPont de Nemours Co., 40, 45, 57
Ekeberg, Anders, 20
elements: essential to life, 21–22; history of use, 15–21; in man, 6–14; in seawater, 27–29; toxicity of, 3–5, 24–25
emission control devices, 55–58
emphysema, 84
"empty" calories, 120–24
environmental contamination, 26–35
Environmental Defense Fund, 39
Environmental Protection Agency (EPA): air quality standards of, 132; and cadmium in fish, 75, 77; and lead pollution data, 44–47; and leaded gasoline, 54–55, 58; NTA banned by, 84
EPA. *See* Environmental Protection Agency
Ethyl Corporation, 40–41, 45, 48
evolution, 9–13

fat: in processed foods, 119–20; refined, 125
FDA. *See* Food and Drug Administration
Federal Register, 55
feed additives, 100, 102
Finch, Robert, 39
fish: cadmium in, 75–79; mercury in, 63–72
fluoridation of water supply, 23
food, 111–29: cadmium in, 90–91; empty

calories in, 120–24; micronutrients in, 111–29; nutritional situation in U.S., 112–17; recommended daily allowances, 128; trace elements in hospital diet, 118
Food Additives Act, 100
Food and Drug Administration: and arsenic, 98, 101; and labeling of packaged food, 127; and lead, 43; and mercury in fish, 63, 70, 71; and selenium additives, 102; and zinc, 115
Food and Nutrition Board, 113–15, 124–27
fossil fuels: air contamination by burning of, 28–31; cost of burning, 107, 109–10
Foundry Cove, 76–77

gallium, 105
gasoline, leaded, 37, 55–58, 132
gasoline shortage, 58
germanium, 18–19, 105
Gillette, Robert, 47
Goldsmith, John R., 38
Gonzales, Thomas Arthur, 74
Great American Diet, 113–29

hair, arsenic in, 99–101
hardening of the arteries, 125–27
Hardy, Harriet L., 41, 45, 46
Hart, Philip A., 54, 76
Hatchett, Charles, 20
health: diet for, 111–29; elements essential to, 22
heart disease: and cadmium, 87–88; and "water factor," 130–31
high blood pressure, 81–88
history of man's use of metals, 15–21
homeostatic mechanisms, 13–14
hospital diet, 117–18
Hudson River pollution, 74–78
human body: elements in, 21–24; exposures to toxic metals, 93–94, 133–34; mercury in, 63; metals in, 6–14
hypertension, 81–88

ice trays, cadmium on, 74
industry: cost of pollution by, 107, 109–10; elements consumed by, 21–22, 24; environmental contamination by, 26–35; exposures of workers in, 133–34; lead, 36–58
insecticides, arsenic in, 99
itai-itai disease, 77, 83, 92

Japan, cadmium-caused disease in, 77, 83, 92
Justice Department, 75–77

Kehoe, Robert A., 38, 45, 46
Kettering Laboratory, 40, 41
kidneys, cadmium in human, 91–92
Kretchmer, Jerome, 58

lakes, metals in, 76
lanthanum, 105
lead, 36–58: airborne in cities, 40–42;
    control of toxicity of, 132; overex-
    posure to, 56; history of use, 17; as
    pollution source, 51; sources of, 57;
    in street dirt, 55
Lead Industries Association (LIA), 37,
    43–44, 53, 55
leaded gasoline, 37, 55–58, 132
LIA, 37, 43–44, 53, 55
life: elements essential to, 22; evolution
    of, 9–13
livers, cadmium in human, 91–92
lung cancer from beryllium, 95–96

man: elements in, 21–24; evolution of,
    9–13; mercury in, 63; trace elements
    in, 6–14; *see also* human body
manganese, 20–21, 116
Marathon Battery Company, 76
Marshall, Louise, 46
mercury, 59–72: industrial discharges
    of, 65–66; methylation of, 67–72, 84–
    86; in nature, 62; pollution control,
    131–32; sources of exposure to, 61
metals: cadmium, 73–92; contamination
    by, 26–35; control of pollution by,
    130–37; diseases of workers exposed
    to, 106–7; essential to life, 21–22; in
    food, 111–129; history of use, 15–21;
    in human body, 6–14; human expo-
    sures to, 94, 106–7; in human hair,
    101; industrial, 18; lead, 36–58; mer-
    cury, 59–72; in rivers and lakes, 76;
    toxicity of, 5, 24–25, 93–109
methyl mercury, 67–72, 84–85
methylation, 67–72
micronutrients in food, 111–29
milk, fat-free, 119–22
Mitre Corporation, 132
molybdenum, 116

Nader, Ralph, 113
Napoleon, 99–100
Nason, Alexis P., 76–77
National Academy of Sciences, 44–47
National Air Pollution Control Admin-
    istration, 47
National Broadcasting Company, 43
National Research Council, 38, 39, 44–
    47, 113
Natural History, 36

nature, contamination by, 26–35
NBC, 43
nickel, 104–5, 132
niobium, 20
NRC, 38, 39, 44–47, 113
NTA, 84–85
nutritional content of food, 111–29

ocean, pollution of, 26–33, 105–7; *see
    also* seawater
ouch-ouch disease, 77, 83, 92

Patterson, Clair C., 40, 48
pesticides, arsenic in, 99
plant evolution, 9
poisoning, lead, 36–58; *see also* toxicity
pollution: of air, 105–10; control, 130–
    37; cost of, 107, 109–10; defined, 4;
    by industry, 107–10; by nature, 26–
    35; of ocean, 105–7; by weathering,
    107, 109–10; *see also* contamination
processing of food, 119–29
Proust, J. L., 70
Prouty, Winston, 65
Public Health Service, 84–85

radioactivity, 21
Recommended Daily Allowances (RDA),
    127–28
Reference Man: elements in, 21–24;
    mercury in, 63; *see also* human body
refrigerator trays, cadmium on, 74
refining of foods, 120–29
re-refining of petroleum, 49
rhodium, 105
Rickover, Hyman, 39
river pollution, 74–78
Rose, Heinrich, 20
Ruckelshaus, William P., 54

salt water, 11–12; *see also* seawater
Schroeder, Henry A., 38
Science, 44
seawater: elements in, 21–24, 27–29;
    metals added to by weathering, 109–
    10; potential pollution of, 31–32, 105–
    7; salinity of, 11–12; toxicity of, 29–
    31
selenate, 102
selenite, 102
selenium, 19, 101–3
Seymour, David M., 76
shellfish, cadmium in, 78–79
skeletons, 10–11
Sonotone Corporation, 76
Stigler, R., 98
Stock, A., 70
Stopps, Gordon J., 45, 46

street dirt, lead in, 55
sugar, 120–21, 125

tantalum, 20
tin: in food containers, 112; toxicity of, 104–5
Tipton, Isabel H., 41, 74
toxicity: of elements, 3–5, 24–25; potential human exposures to, 93–94
Trace Element Laboratory of Dartmouth College, 38, 52, 70, 75, 102
trace elements: in food, 111–29; in man, 6–14; toxic, 5
tungsten, 105

U.S. agencies. *See under name of agency*, i.e., U.S. Public Health Service. *See* Public Health Service

vanadium, 20
Vermont: environmental control in, 134; game fish in, 70–72

vitamins: B₁₂, 117; recommended daily allowances of, 128
volcanic pollution, 26–27

water: drinking, and heart disease, 88, 130–31; drinking, standards for, 134–36; potential pollution of, 31–32; salt, 11–12; standards for, 134–36; *see also* seawater
weathering, cost of, 107, 109–10
Westwöö, Gunnel, 68
Winn, Ira J., 37, 39, 53
Winthrop, John, 20
workers: diseases of, 106–7; exposure to cadmium, 91–92; exposure to pollution, 133–34
World Health Organization, 39
Worthington, V. T., 49

zinc: and cadmium poisoning, 77, 80, 86; deficiency of, 115
zirconium, 105